new matrix   1

The Biology
of Ultimate
Concern

new matrix

The Biology
of Ultimate
Concern

Theodosius Dobzhansky

© 1967 THEODOSIUS DOBZHANSKY
FIRST PUBLISHED IN GREAT BRITAIN 1969 BY
RAPP AND WHITING LIMITED
76 NEW OXFORD STREET LONDON WC1
PUBLISHED IN THE UNITED STATES OF AMERICA BY
THE NEW AMERICAN LIBRARY INC NEW YORK
PRINTED IN GREAT BRITAIN BY
CLARKE DOBLE AND BRENDON LIMITED AT PLYMOUTH

Contents

Preface ad
Hominem

The *Weltanschauung,* some
components of which are sketched in the pages that follow,
was nurtured, modified, and corrected for something close
to half a century. Its germs arose when the author was in his
teens, and became naively enraptured with evolutionary
biology. The intellectual stimulation derived from the
works of Darwin and other evolutionists was pitted against
that arising from reading Dostoevsky, to a lesser extent
Tolstoy, and philosophers such as Soloviev and Bergson.
Some sort of reconciliation or harmonization seemed necessary. The urgency of finding a meaning of life grew in the
bloody tumult of the Russian Revolution, when life became most insecure and its sense least intelligible.

Among my teachers and mentors I was lucky to have, in
Russia, Professors S. E. Kushakevich, V. I. Vernadsky, later
J. A. Philipchenko, and in America, T. H. Morgan. I have
profited by many discussions and debates with my friends
and colleagues S. I. Obolensky, A. P. Obnovlensky, and
M. M. Levit in Russia; L. C. Dunn, A. E. Mirsky, G. G.
Simpson, and E. Mayr in America; and L. C. Birch in
America, Brazil, and Australia. This acknowledgment of
intellectual debts does not at all imply that the persons
mentioned agree with the ideas expressed in this book. I
have met P. Pierre Teilhard de Chardin on two occasions,
but failed to appreciate his greatness until the posthumous
publication of his books. The works of A. J. Toynbee and

Paul Tillich have influenced my thinking. During the Third Delos Symposium, in July 1965, I had the privilege of meeting Dr. Toynbee and of discussing with him some parts of this book, which were at that time in first draft. Whatever expertness I may possess is in biology, more precisely in evolutionary genetics. This is no warrant for embarking on speculations in the realms of philosophy and religion. Such speculations are often regarded, among scientists, as regrettable foibles or even as professional misdemeanors. They are as often as not kept secret, for being caught at them is liable to damage a scientist's professional reputation. Let me, then, try to make clear the nature of my enterprise. This is not an attempt to derive a philosophy from biology, but rather to include biology in a *Weltanschauung*. It will probably be fairly generally admitted that biology is relevant to philosophy. Perhaps it is even more relevant than most other sciences. A biologist, qua biologist, may therefore be competent to sort out and to present facts, theories, and ideas which he believes to be of general humanistic interest and import.

I must acknowledge the kindness of Drs. S. D. Coe, M. D. Coe, F. Ayala, and L. C. Birch, who have read and criticized the manuscript of this book. I have profited by their remarks, but this does not make them in the least responsible for any errors of commission or omission, and does not imply agreement with all ideas and points of view in the book, for which I am solely responsible.

T. D.

I

Humanism
and Humanity

Dostoevsky makes his Ivan Kara-mazov declare: "What is strange, what is marvelous, is not that God really exists, the marvel is that such an idea, the idea of the necessity of God, could have entered the head of such a savage and vicious beast as man; so holy it is, so moving, so wise, and such a great honor it does to man." This is even more marvelous than Dostoevsky knew. Mankind, *Homo sapiens,* man the wise, arose from ancestors who were not men, and were not wise in the sense man can be. Man has ascended to his present estate from one still more savage, not necessarily more vicious, but quite certainly a dumb and irrational one. It is unfortunate that Darwin has entitled one of his two greatest books the "Descent," rather than the "Ascent," of man. The idea of the necessity of God, and other thoughts and ideas that do honor to man, were alien to our remote ancestors. They arose and developed, and secured a firm hold on man's creative thought during mankind's long and toilsome ascent from animality to humanity.

Organisms other than men have the "wisdom of the body"; man has in addition the wisdom of humanity. Wisdom of the body is the ability of a living system so to react to environmental changes that the probabilities of survival and reproduction are maximized. For example, a certain concentration of salt in the blood is necessary for life; if an excess of salt is ingested, it is eliminated in the urine; if the salt supply is scarce the urine contains little salt. Such "wise" reactions of the body are confined usually to the environments which the species has frequently encountered in its evolutionary development. This built-in "wisdom" arose through the action of natural selection.

The place of the wisdom of humanity in the scheme of things requires a separate consideration. Humanism, according to Tillich (1963), "asserts that the aim of culture is the actualization of the potentialities of man as the bearer of spirit," and "Wisdom can be distinguished from objectifying knowledge (*sapientia* from *scientia*) by its ability to manifest itself beyond the cleavage of subject and object." This wisdom is a fruit of self-awareness; man can transcend himself, and see himself as an object among other objects. He has attained the status of a person in the existential sense, and with it a poignant experience of freedom, of being able to contrive and to plan actions, and to execute his plans or to leave them in abeyance. Through freedom, he gains a knowledge of good and of evil. This knowledge is a heavy load to carry, a load of which organisms other than man are free. Man's freedom leads him to ask what Brinton (1953) refers to as Big Questions, which no animals can ask.

Does my life and the lives of other people have any meaning? Does the world into which I am cast without my consent have any meaning? There are no final answers to these Big Questions, and probably there never will be any, if by answers one means precise, objective, provable certitudes. And yet seek for some sort of answers we must, because it is the highest glory of man's humanity that he is capable of searching for his own meaning and for the meaning of the Cosmos. An urge to devise answers to such "metaphysical" questions is a part of the psychological equipment of the human species. Brinton (1953) rightly says that "Metaphysics is a human drive or appetite, and to ask men to do without metaphysics is as pointless as to ask them to do without sex relations. There are indeed individuals who can practice abstention from metaphysics as there are those who can practice abstention in matters of sex, but they are the exceptions. And as some who repress sex actually divert it into unprofitable channels, so do those who repress metaphysics."

The German word *Weltanschauung* and the Russian *mirovozzrenie* have no precise English equivalents. The usual translation, "world view," subtly betrays the meaning. A world view, like a view from a mountaintop, may be pleasant and even inspiring to behold, but one can live without it. There is a greater urgency about a *Weltanschauung*, and some sort of *mirovozzrenie* is felt to be indispensable for a human being. The Latin *credo* is becoming acclimatized in English in a sense most nearly equivalent to *Weltanschauung*. It is most closely related to the "ultimate concern" which Tillich considers to be the essence of religion in the broadest and most inclusive sense. "Religion is the aspect of depth in the totality of the human spirit. What does the metaphor depth mean? It means that the religious aspect points to that which is ultimate, infinite, unconditional, in man's spiritual life. Religion, in the largest and most basic sense of the word, is ultimate concern. And ultimate concern is manifest in all creative functions of the human spirit" (Tillich 1959).

It is the ultimate concern in man that Ivan Karamazov found so strange and so marvelous. Man's nature impels him to ask the Big Questions. Every individual makes some attempts to answer them at least to his own satisfaction. One of the possible answers may be that the Questions are unanswerable, and only inordinately conceited or foolish people can claim to have discovered unconditionally and permanently valid answers. Every generation must try to arrive at answers which fit its particular experience; within a generation, individuals who have lived through different experiences may, not quite, one hopes, in vain, make sense of those aspects of the world which the individual has observed from his particular situation.

My life has been devoted to working in science, particularly in evolutionary biology. Scientists are not necessarily more, but I hope also not less, qualified to think or to write about the Big Questions than are nonscientists. It is naive to think that a coherent credo can be derived from science alone, or that what one may learn about evolution will unambiguously answer the Big Questions. Some thinkers, e.g., Barzun (1964) dismiss such pretensions with undisguised scorn: ". . . the scientific profession does not constitute an elite, intellectual or other. The chances are that 'the scientist,' from the high-school teacher of science to the head of a research institute, is a person of but average capacity." And yet even Barzun, no friend or respecter of science, grudgingly admits that science "brings men together in an unexampled way on statements to which they agree without the need of persuasion; for as soon as they understand, they concur." Some of these "statements" which science produces are at least relevant to the Big Questions, and in groping for tentative answers they ought not be ignored.

The time is not long past when almost everybody thought that the earth was flat, and that diseases were caused by evil spirits. At present quite different views are fairly generally accepted. The earth is a sphere rotating on its axis and around the sun, and diseases are brought about by a variety of parasites and other biological causes. This has influenced people's attitudes; the cosmology that one credits is not irrelevant to one's ultimate concern. To Newton and to those who followed him the world was a grand and sublime contrivance, which operates unerringly and in accord with precise and immutable laws. Newton accepted, however, Bishop Ussher's calculations, which alleged that the world was created in 4004 B.C. The world was, consequently, not very old; it had not changed appreciably since its origin, and it was not expected to change radically in the future, until it ended in the apocalyptic catastrophe. Newton was a student of the Book of Revelation as well as a student of cosmology. In Newton's world man had neither power enough nor time enough to alter the course of events which were predestined from the beginning of the world.

The vast universe discovered by Copernicus, Kepler, Galileo, and Newton became quite unlike the cozy geo-centric world of the ancient and the medieval thinkers. Man and the earth were demoted from being the center of the universe to an utterly insignificant speck of dust lost in the cosmic spaces. The comfortable certainties of the traditional medieval world were thus taken away from man. Long before the modern existentialists made estrangement and anxiety fashionable as the foundations of their phi-losophies, Pascal expressed most poignantly the loneliness which man began to feel in "The eternal silence of these infinite spaces." If he worked hard, man could conceivably learn much about how the world was built and operated, but he could not hope to change it, except in petty detail. An individual human was either saved or damned, and those of Calvinist persuasion believed that this alternative was irrevocably settled before a person was even born. This left no place for humanism in Tillich's sense; an individual man had few potentialities to be actualized, and culture had scarcely any at all.

It has become almost a commonplace that Darwin's dis-covery of biological evolution completed the downgrading and estrangement of man begun by Copernicus and Gali-leo. I can scarcely imagine a judgment more mistaken. Perhaps the central point to be argued in this book is that the opposite is true. Evolution is a source of hope for man. To be sure, modern evolutionism has not restored the earth to the position of the center of the universe. However, while the universe is surely not geocentric, it may con-ceivably be anthropocentric. Man, this mysterious product of the world's evolution, may also be its protagonist, and eventually its pilot. In any case, the world is not fixed, not finished, and not unchangeable. Everything in it is engaged in evolutionary flow and development.

Human society and culture, mankind itself, the living world, the terrestrial globe, the solar system, and even the "indivisible" atoms arose from ancestral states which were radically different from the present states. Moreover, the changes are not all past history. The world has not only evolved, it is evolving. Now, "In the Renaissance view, the world, a place of beauty and delight, needed not to be changed but only to be embraced; and the world's people, free of guilt, might be simply and candidly loved" (Dunham 1964). Far more often, it has been felt that changes are needed:

For the created universe waits with eager expectation for God's sons to be revealed. It was made the victim of frustration, not by its own choice, but because of him who made it so; yet always there was hope, because the universe itself is to be freed from the shackles of mortality and enter upon the liberty and splendor of the children of God. Up to the present, we know, the whole created universe groans in all its parts as if in the pangs of childbirth (Rom. 8 : 19–22).

Since the world is evolving it may in time become different from what it is. And if so, man may help to channel the changes in a direction which he deems desirable and good. With an optimism characteristic of the age in which he lived, Thomas Jefferson thought that "Although I do not, with some enthusiasts, believe that the human condition will ever advance to such a state of perfection as that there shall no longer be pain or vice in the world, yet I believe it susceptible of much improvement, and most of all, in matters of government and religion; and that the diffusion of knowledge among the people is to be the instrument by which it is to be effected." This is echoed and reechoed by Karl Marx and by Lenin in their famous maxim that we must strive not merely to know but also to transform the world. In particular, it is not true that human nature does not change; this "nature" is not a status but a process. The potentialities of man's development are far from exhausted, either biologically or culturally. Man must develop as the bearer of spirit and of ultimate concern. Together with Nietzsche we may say: "Man is something that must be overcome."

Picasso is alleged to have said that he detests nature. Tolstoy and some lesser lights claimed that any and all findings of science made no difference to them. Fondness and aversion are emotions which admittedly cannot be either forcibly implanted or expurgated. One may detest nature and despise science, but it becomes more and more difficult to ignore them. Science in the modern world is not an entertainment for some devotees. It is on the way to becoming everybody's business. Some people feel no interest in distant galaxies, in foreign lands, exotic human tribes, and even in those neighbors with whom they are not constrained to deal too often or too closely. Indifference to one's own person is unlikely. It is feigned by some, but rarely felt deep down, when one is all alone with oneself. This unlikelihood, too, is understandable as a product of the biological evolution of personality in our ancestors. It made the probability of their survival greater than it would have been otherwise. Ingrained in man's psyche before it was explicitly formulated, the adage "Know thyself" was always a stimulus for human intellect.

To "know thyself," scientific knowledge alone is palpably insufficient. This was probably the basis of Tolstoy's scoffing at science. To him science seemed irrelevant to the ultimate concern, and to him only the ultimate concern seemed to matter. But he went too far in his protest. In his day, and far more so in ours, the self-knowledge lacks something very pertinent to the present condition if one chooses to ignore what one can learn about oneself from science. This adds up to something pretty simple, after all: a coherent credo can neither be derived from science nor arrived at without science.

Construction and critical examination of credos fall traditionally in the province of philosophy. Understandably enough, professional philosophers often show little patience with amateurs who intrude into their territory. Scientists turned philosophers fare scarcely better than other amateur intruders. This proprietary attitude is not without warrant, but the matter is not settled quite so easily. What, indeed, is philosophy? Among the numerous definitions, that given by Bertrand Russell (1945) is interesting: "between theology and science there is a No Man's Land, exposed to attacks from both sides; this No Man's Land is philosophy." Less colorfully, philosophy is defined as the "science of the whole," which critically examines the assumptions and the findings of all other sciences, and considers them in their interrelations. Still other definitions claim that philosophy works to construct a coherent *Weltanschauung*. Under any of these definitions, scientists may have some role to play, at least on the outskirts of philosophy. At the very least, they must be counted among the purveyors of raw materials with which philosophers operate when they formulate and try to solve their problems. With some notable exceptions, modern schools of philosophy, especially in the United States and in England, have been taking their cues very largely from the physical sciences; the influential school of analytical philosophy is engrossed with mathematics and linguistics. Biology and anthropology are neglected. Of late, there appear to be, however, some straws in the wind portending change.

The relevance of biology and anthropology is evident enough. In his pride, man hopes to become a demigod. But he still is, and probably will remain, in goodly part a biological species. His past, all his antecedents, are biological. To understand himself he must know whence he came and what guided him on his way. To plan his future, both as an individual and much more so as a species, he must know his potentialities and his limitations. These problems are only partly biological and scientific, and partly "theological." In short, they are philosophical problems in Bertrand Russell's sense.

Since I am a biologist without formal philosophical or anthropological training, the task which I set for myself is quite likely overambitious. I wish to examine some philosophical implications of certain biological and anthropological findings and theories. This small book lays no claim to being a treatise either on philosophical biology or on biological philosophy. It consists of essays on those particular aspects of science which have been particularly influential in the formation of my personal credo. This is said not in order to disarm the potential critics of these essays, but only to explain what may otherwise appear a rather haphazard selection of topics discussed and of those omitted in the pages that follow. Together with Birch (1965) I submit that:

My scientific colleagues might well say, "Cobbler, stick to your last." But we have been doing that in science for long enough. I have attempted what is not a very popular endeavor in our generation. It is to cover a canvas so broad that the whole cannot possibly be the specialized knowledge of any single person. The attempt may be presumptuous. I have made it because of the urgency that we try, in spite of the vastness of the subject. I would not have written had I not discovered something for myself that makes sense of the world of specialized knowledge in which I live.

2

On Gods
of the Gaps

The experience of discovering or learning a new truth should be, and usually is, exciting and exhilarating. There are, however, rather odd exceptions. Probably most teachers of biology have found a minority of their students perplexed by having what they believe to be "mysteries" of nature explained and clarified. To these students, the ebbing of mystery is a threat to the security of their faiths. To them, there must be gaps between natural events to accommodate God's interventions. Conversely, the cruder forms of antireligious propaganda, in Russia and elsewhere, contend that since science has explained almost everything in the world, there is really nothing left in God's special domain, which must be a realm of mystery and of miracles.

Newton was able to account by physical laws for the motions of the planets around the sun. Nevertheless, he thought that "The diurnal rotations of the planets could not be derived from gravity, but required a divine arm to impress it on them." About a century later, Emperor Napoleon asked the mathematician and cosmologist Laplace why his celebrated work on celestial mechanics contained no mention of God. The Emperor's question was possibly in jest; Laplace replied in all seriousness that in his work he had "no need of that hypothesis." Laplace did not say so, but many people concluded that if God is not needed in celestial affairs, there should be even less need of him in terrestrial ones. Countless books and essays in many languages have been written about the unremitting decline of the utility of the God "hypothesis" for scientific explanation of the observable world. A. D. White's *Warfare of Science and Theology* (1895) remains a classic of this genre.

The French revolutionary tribunal, which condemned Lavoisier to die, declared that "la République n'a pas besoin de savants." If almost everything in the world were already known and explained, this rather unenlightened opinion would perhaps not be so far from the truth. The work of research scientists would then be rather boring, and scientists as an occupational group would be superfluous. This is, however, far from true; the enterprise of science is still in its beginning stages. Only some phenomena of nature are in part understood, and the realm awaiting explanation is far greater. It is precisely the unknown that inspires scientists to press on their quest. There are people, however, to whom the gaps in our understanding of nature are pleasing for a different reason. These people hope that the gaps will be permanent, and that what is unexplained will also remain inexplicable. By a curious twist of reasoning, what is unexplained is then assumed to be the realm of divine activity. The historical odds are all against the "God of the gaps" being able to retain these shelters in perpetuity. There is nothing, however, that can satisfy the type of mind which refuses to accept this testimony of historical experience.

According to Durkheim (1915), "All known religious beliefs, whether simple or complex, present one common characteristic: They presuppose a classification of all things, real and ideal, of which men think, into two classes or opposed groups . . . profane and sacred." This does not quite mean that the profane is the realm of natural and the sacred of supernatural forces. Such a distinction would anyway be meaningless to primitive man, who sees everywhere manifestations of forces that we would consider fanciful or supernatural. Tylor and other nineteenth-century pioneers of cultural anthropology saw the origin of religion in animism. Primitive man supposedly made no distinction between material and spiritual processes. To him, not only other men but also animals, plants, rocks, streams, and even weather, all were animated and had wills of their own. They could choose to be helpful or injurious to man.

These speculations about the origin of religion accorded well with the philosophical doctrines of Comte and Spencer, who assumed three stages of the intellectual development of mankind, from theology, through metaphysics, to science. More recent anthropological work has shown this to be an oversimplification. Malinowski (1931) and others have stressed that scientific and religious ideas have different functions in the social life of people. Malinowski pointed out that "as soon as man developed the mastery of environment by the use of implements, as soon as language came into being, there must also have existed primitive knowledge of an essentially scientific character. No culture could survive if its arts and crafts, its weapons and economic pursuits, were based on mystical, nonempirical conceptions and doctrines." Animism is related to magic, and the function of magic is separate and supplemental to the primitive "science." Man's knowledge is incomplete; so much depends on chance, accident, on forces that are not understood. These forces, says Malinowski, man attempts to control by magic. "Magic is used as something which, over and above man's equipment and his force, helps him to master accident and to ensnare luck." Psychoanalytically oriented authors have arrived at still different conceptions. To Freud, God was not only an unnecessary hypothesis but also a delusional idea stemming from unresolved childhood conflicts. Jung, on the contrary, believed that underlying the experience of a human individual there is an inherited collective unconscious of the human species, which contains "archetypes" expressing themselves in symbolic images (see Munroe 1955, Kardiner 1963).

The full depth of the problem of the applicability of scientific knowledge to the understanding of nature was explored more than three centuries ago by Descartes. The Cartesian solution was a radical dualism. The world is a machine: "I do not accept or desire any other principle in Physics than in Geometry or abstract Mathematics, because all the phenomena of nature may be explained by their means, and sure demonstration can be given of them." Living bodies are likewise machines, and so is the human body. But man is not only a machine; he has in addition a non-mechanical soul. What evidence forced Descartes to make this concession to dualism? Man expresses his thoughts in speech; Descartes (1637) prophetically envisaged that a machine might be constructed that would utter words; however, he thought that such a machine could not "arrange its speech in various ways, in order to reply appropriately to everything that may be said in its presence as even the lowest type of man can do." The Cartesian dualism was obviously unconvincing, and it did not endure. A century after Descartes, La Mettrie (1748), d'Holbach (1770), and Helvetius (1772) saw no reason to ascribe to man something called a soul. Man was to be understood to be as much a machine as any other animal, perhaps slightly more intricate in its construction, to allow for more complex mental processes.

A modern scientist can only marvel at the courage, or temerity, of these "philosophes" of the Age of Enlightenment, who could unabashedly make such sweeping assertions on the basis of evidence so pitifully small compared even to what is available in our day, and what still seems in many ways ambiguous and susceptible to a variety of interpretations. As seen in retrospect, the state of biology in the eighteenth century might have suggested vitalism rather than mechanism as the more reasonable hypothesis. As an illustration, consider just one problem, that of embryonic development. Hamm, a pupil of the pioneer microscopist Leeuwenhoek, discovered human spermatozoa in 1677. In

1694, another microscopist added to this real discovery a spurious one—he claimed that he saw in the head of the spermatozoon a diminutive human figure, a homunculus. He even published a rather imaginative drawing of his supposed discovery.

This made the problem of development seem deceptively simple—if the body is preformed in the sex cell, it needs only to increase in size to become an adult body. The tough problem, how a new body is formed, is conjured away. Even so, this hides a very serious difficulty: where do the next and the following generations come from? A radical solution of the puzzle was the theory of *"emboîtement"*: the homunculus contains minute spermatozoa with even more minute homunculi, these contain still smaller homunculi, etc. All mankind was preformed in the first man (or in the first woman, if the homunculi were contained in the female generative elements). If anybody tried to estimate the sizes of the smallest homunculi, he modestly refrained from publishing the results of his calculations.

The theory of preformation was challenged by champions of epigenesis. Quite rightly, they pointed out that there are no homunculi in the sex cells. The body develops, they thought, from undifferentiated matter, which is certainly an exaggeration. C. F. Wolff showed in 1759 and 1786 that nothing like parts of adult bodies are found either in the early embryos of animals or in the seed primordia. What then are the causes that bring about the development? The answer Wolff offered was not convincing. With the benefit of hindsight, it is easy to scoff at the misconceptions of our predecessors. Wolff and others thought that there must be a vital force, a *vis essentialis*, which directs the embryonic development and subsequently maintains the life of the adult body. Many eighteenth-century biologists, including Wolff, John Hunter, Stahl, Bichat, were vitalists. Before a reasonable picture of embryonic development could be obtained, several generations of embryologists had to betake themselves to their dissecting instruments and their progressively improving microscopes, and to make painstaking descriptions of just how the representatives of different groups of the animal and plant kingdoms are formed from fertilized egg cells, through different stages of their life cycles, to the production of new generations.

The theory of germ layers in animals was developed by von Baer (1828, 1837). The Schleiden (1838) and Schwann (1839) cell doctrine made the description of the development easier and more exact. Haeckel (1866) formulated the "biogenetic law"—the individual development, ontogeny, is a short recapitulation of the evolutionary development, phylogeny, of the group to which the developing organism belongs. This proved to be an example of a scientific theory which contained only a partial truth, but which played a very useful role, since it stimulated a tremendous interest and activity among investigators.

The entry of the spermatozoon into the egg cell and the fusion of their nuclei were described by Hertwig (1875) and Fol (1879). Toward the close of the nineteenth and the beginning of the twentieth century, a brilliant pleiad of scientists described and analyzed the behavior of the chromosomes in cell division (mitosis), in the maturation of the sex cells (meiosis), in fertilization, and in cleavage of fertilized eggs.

This happened to coincide with the rediscovery in 1900 of Mendel's laws (originally reported by Mendel in 1865 but ignored until 1900). There ensued a rapid growth of genetics, one branch of which, developmental genetics, is concerned with analysis of the realization of the hereditary endowment of the organism in the ontogeny. The old contrast of preformation versus epigenesis melted away, or at any rate transformed itself into a different problem. There are no homunculi in the sex cells; there are sets of genes which contain, to use a modern simile borrowed from cybernetics, the "information" or "instructions" for the development. The problem now became to find out how this information is "coded" in the genes. Hypotheses with a preformist flavor postulated a one-to-one relationship between each gene and a part or a "character" of the adult organism, or between the gene and the enzyme which it produces. As seen at present, the real situation appears to be, in a way, even simpler. The Watson-Crick hypothesis, now regarded as amply confirmed by facts, is that heredity is coded in the chainlike molecules of DNA (deoxyribonucleic acid). Different genes are comparable to words or paragraphs on a printed page—they are different sequences or combinations of four nucleotides, the four "genetic letters." The genetic messages coded in the DNA of the chromosomal genes are "transcribed" into a still different code in ribonucleic acids (RNA), and into sequences of amino acids which form parts of protein molecules. Some proteins act as enzymes.

All this surely has not led to a Cartesian reduction of embryology and genetics to mechanics, geometry, and mathematics. On the other hand, nothing at all has been discovered that would necessitate, or even suggest, the assumption of any kind of vital force. Vitalism in embryology and genetics has gradually withered away; the vital force, the existence of which seemed so certain to the pioneer students of development has nowhere been found. Driesch (1867–1941) was perhaps the last eminent embryologist who acknowledged being a vitalist. Around the turn of the century Driesch made ingenious experiments on the development of sea urchins. The embryos of these animals possess a remarkable capacity of regeneration of missing parts; isolated cells may develop into diminutive but fairly complete larvae. This, we would think at present, is due to the cells' preserving the full complement of the genes which the egg cell also has. But Driesch thought otherwise. To him, no machine could ever be so constructed that it would restore the missing parts. The development must be guided by something for which Driesch borrowed Aristotle's word "entelechy." But what is an entelechy? Driesch thought that entelechy is located nowhere spatially or anatomically; it is neither a special substance nor a new kind of energy; it is "the elemental teleological factor of nature." Most biologists thought that it was just a word, explaining nothing.

The situation in other branches of biological science is rather similar to that outlined above. Digestion, respiration, hormone action, muscle contraction, nerve conduction, and finally the structure and operation of the brain, all present innumerable problems which have been studied by several generations of biologists. The hypothesis of mechanism has triumphed everywhere, if you wish, by default; whatever biological process has been successfully analyzed has proved to represent patterns, usually exceedingly complex ones, of chemical and physical components. It is this remarkable complexity, not the presence of some peculiar vital forces, that constitutes the core problem of biology. As to the mechanism/vitalism contest, it has been pretty nearly a dead issue in biology for about half a century.

Every student taking an introductory biology course is expected to learn that mechanism is right and vitalism is wrong. However, not every student, and not even every professor, has a clear understanding of why this is so. Biology is very far from having accomplished the Cartesian program of reducing all biological phenomena to chemistry, physics, and mathematics. No sensible mechanist expects to accomplish anything of the sort. This would be like proving a universal negation; there are too many different organisms with countless different structures or functions. One can always imagine a vital force lurking in some place not reached by study. Vitalism is rejected for a different reason. The all but unanimous consensus is that vitalism is useless as a working hypothesis in biological research. It can no longer serve even as a goad to nonvitalists to vindicate their doctrines.

When Wöhler, in 1828, synthesized urea, the challenge

was to prove that a substance hitherto found only in living bodies can also be produced outside them. Nowadays, synthesis of a new organic compound is no longer news, unless the particular compound is interesting for some special reason. To do biological or biochemical research for the purpose of obtaining further evidence against vitalism makes as much sense as fighting windmills. Tremendous strides have been made in recent years in molecular biology and biochemistry. Molecular biology is Cartesian in its inspiration. Its chief aim is, however, no longer to test the organism-the-machine theory. The discovery of the structure of the nucleic acids, breaking the "genetic code," unraveling the mechanisms by which the molecules that bear the genetic information reproduce themselves, are among the greatest achievements of modern science. They are important because they help us to understand the workings of heredity; they were not needed to convince biologists that genes operate through chemical messengers, rather than through exertions of a vital force. Existence of some kinds of chemical messengers was already considered overwhelmingly probable by T. H. Morgan, R. B. Goldschmidt, and other geneticists several decades before the DNA era. It would be hard to imagine how else the genes could possibly act.

Scientific problems should not, however, be settled by majority votes. The notions of the heterodox minority who find biology too crudely materialistic for their taste deserve a fair hearing. In the face of the splendid achievements of mechanist science, this minority see biology in need of vitalist reorientation. What characterizes the situation better than anything else is that modern vitalists like to avoid being called vitalists. The ugly word is rarely mentioned in Sinnott's eloquent arguments (1955, 1957). The following quotation gives, I think, a fair idea of his basic biological conviction: "Organized system, maintained by the regulatory control of its activities, implies the presence within it of *something to which these activities tend to conform,* a norm, a standard, a goal or end, what the philosopher would call a *telos,* inherent in the whole living mass" (author's emphasis). Being a distinguished biologist, Sinnott is at home with the present state as well as the history of biology. Yet his biological argumentation is essentially the same as that given years ago by Driesch. It can be summarized in two affirmations. First, the development of an organism is goal-directed, the goal being the attainment of the adult state. Second, the physiology and behavior of an organism are also goal-directed, the goal being the continuation of life.

It is possible to accept Sinnott's facts, though not his conclusions. Although some purists take exception to the statement that organized systems have goals or ends, I see no need to quibble about these words. Processes which are observed in a developing embryo make sense only when one knows what develops from this embryo. Almost everything which an organism does, physiologically and behaviorally, serves to enable this organism to stay alive and to resist destruction (for an excellent discussion, see Simpson 1964). The real issue is whether this proves, or even makes plausible, the assumption of a "something" called "telos" or vital force. It does not, in my opinion. The undoubted fact that organisms are adapted to the environments in which they live is understandable in the light of the theory of evolution. An "organized system, maintained by the regulatory control of its activities" implies not a "telos" but a history of evolutionary adjustment. This will be discussed, of necessity very briefly, in the next chapter. Here attention should be called to the function which evolutionary explanations must generally play in biology. It is an underestimation of the significance of the evolutionary origin of organisms that lies, I believe, at the base of the neovitalist misconception.

At a risk of oversimplifying a situation that is far from simple, it can be stated that the study of the living world, including man, is approached in two ways. Both are equally legitimate and necessary, but they ask somewhat different questions, and therefore naturally come up with different, though complementary, answers. The first is the reductionist, or Cartesian, approach. The components of the living bodies, ultimately the chemical substances of which these bodies are built, are examined and analyzed. Stated broadly, the explanations sought are of the organism-the-machine type. In the words of Wald (1963), "Living organisms are the greatly magnified expressions of the molecules that compose them." The other approach is Darwinian. It is more specifically, though not exclusively, biological. We inquire how living bodies got to be built as they are. Life and living beings can be made intelligible in the light of their origins and of their phylogenetic development, i.e., in the light of evolution. Simpson (1964) has stated this as follows:

In biology, then, a second kind of explanation must be added to the first or reductionist explanation made in terms of physical, chemical, and mechanical principles. This second form of explanation, which can be called compositionist in contrast with reductionist, is in terms of the adaptive usefulness of structures and processes to the whole organism and to the species of which it is a part, and still further, in terms of ecological function in the communities in which the species occurs.

By and large, the closer one works to the chemical level the more relevant and more nearly complete will explanations of the Cartesian type be. The need of Darwinian approaches increases as one moves upwards to the organismic levels. Let me stress, however, that there is no sharp dichotomy. Cartesian as well as Darwinian explanations are meaningful on all biological levels, although they look at the problem of life from different angles. They are not rival or competing but complementary.

Consider this miracle, which astounds nobody because it is an everyday occurrence—the coming into the world of a new human life. A human egg cell weighs about one twenty-millionth of an ounce. A grown-up is some fifty billion times heavier. Where does the material for this growth come from? It is derived obviously from the environment, from the food consumed and assimilated. Quite literally, man is a conglomeration of transformed groceries. This conglomeration is, however, alive, feels joy and suffering, possesses consciousness and self-awareness, and is capable of deeds ranging from kindness to knavery and from heroism to egoism. The idea that a conglomeration of groceries, however transformed, can do any of these things strikes some people as arrant nonsense. When it is seen as inescapable it becomes a source of despair. Thus to Barzun (1964):

All this cannot help fostering the conviction that no conceivable life is worth living. The more heart and mind a person has, the readier he is to feel the force of these fears, commands and privations, and to wish for nothing but an early end: there is clearly no use struggling to save a world in which one would find it hell to live when saved.

To Matson (1964), science makes man "The Alienated Machine."

Sinnott's vitalism is an attempt to escape the despair. "Or is it not possible," asks Sinnott (1957), "—and this is the alternative that here has been proposed—that there is a *natural, inherent directiveness* in living stuff which, subject to modification by factors inside the body and outside, expresses itself in what a person thinks and does and is? Is not life itself, at least to a degree, *creative?*" (author's emphasis). And also: "If the suggestion here presented is correct, this quality is part of the universal organizing, creative power in nature which may be thought of as an attribute of God." Sinnott evidently feels a need to have gaps between natural events to accommodate his God.

I wish to suggest that there is a third way, which avoids

both Barzun's despair and Sinnott's anachronistic vitalism. To describe man as a conglomeration of transformed groceries is valid and realistic, up to a certain point. The limit of the validity of this description is set by its reductionist character. What is needed in addition is a compositionist counterpart. Man certainly consists of molecules and atoms, but he does not arise by an accidental concourse of these molecules and atoms. The fact that must constantly be kept before our eyes is that every organism of any species alive today is a direct lineal descendant of some kind of primordial life, which is estimated to have appeared two or more billion years ago.

Every generation involves formation and dissolution of individuals, which from a reductionist standpoint are patterns of atoms and molecules, all derived from "groceries," i.e., from food. But the history of life is not like Nietzschean eternal recurrence. The patterns constructed in successive generations are not exact repetitions. Neither are the generations wholly unlike each other. Heredity makes a descendant generation in general similar to the parental one, but it does not enforce complete identity. Similarities and differences are equally significant. The hereditary continuity makes organisms time-binding contrivances; the adaptations achieved in the past are not easily squandered. And yet, variation permits the accumulation of new information about the environments in which the species lives at present. This information is ordered and stored in the genes by means of natural selection; natural selection is a cybernetic process which responds to the challenges of the environment and transfers the responses to the genetic endowment of the species.

An individual begins its existence as a fertilized ovum, and proceeds to develop through a complex series of maneuvers. Body structures and functions that are formed fit together not because they are contrived by some inherent directiveness named "telos," but because the development of an individual is a part of the cyclic (or, more precisely, spiral) sequence of the developments of the ancestors. Individual development seems to be attracted by its end rather than impelled by its beginning; organs in a developing individual are formed for future uses because in the evolution they were formed for contemporaneous utility. Individual development is understandable as a part of the evolutionary development of the species, not the other way around. The history of the living world has not been wasted; atoms, molecules, and "groceries" achieve in living organisms feats of virtuosity, because natural selection is a process which makes possible the realization of what would be in the highest degree improbable without it.

Rejecting vitalism in no way conflicts with what Albert Schweitzer has called "reverence for life." Man's conscience, the existence of life, and indeed of the universe itself, all are parts of the *mysterium tremendum*. Trying to find gaps in scientific knowledge, which is of course easy to do, turns out not to be the wisest way to approach the *mysterium*. To Coulson (1955),

There seems to be only one way out of our dilemma. If we cannot bring God in at the end of science, he must be there at the very start, and right through it.

The theologian Heim (1953) is no less explicit:

For faith gives us the strength which we need in everyday life, not when it is sustained by miraculous occurrences breaking through the order of nature . . . but only when one and the same occurrence, an occurrence of which we fully understand the natural causes . . . at the same time in itself appears to us as an act of God, which we receive directly from his hands.

An entelechy or a telos would be a sham solution of biological riddles anyway. Consider how an immaterial "elementary teleological factor," as Driesch called it, could conceivably operate to direct the development of an embryo, or to modify adaptively the adult organism. As stated above, Driesch insisted that the entelechy was neither a special substance nor a new kind of energy, and Sinnott, although he is cautiously vague about this point, seems to hold a similar view. A body, or a brain, is, however, admittedly, a material system—this is not doubted even by vitalists. An immaterial or, indeed, supernatural telos must eventually somehow stoop down to the material and natural levels if it is to produce effects observable on these levels. Will the entelechy turn out to be some very powerful enzyme or a novel sort of radiation? Enzymes and radiations are, however, substances or energies.

Suppose, for the sake of argument, that the parapsychologists adduced incontrovertible proofs of the reality of telepathy or of mediumistic phenomena. Would this bear out vitalism as a biological theory? These are the phenomena which most people would think of as examples of the immaterial or supernatural. But are they? Telepathy could be imagined as being transmitted by some not hitherto known force, but whatever that force may be it must necessarily be able to act on the nerve cells of the receiver. The new form of energy must cause in these cells some physiological biochemical processes, which will later be translated, also physiologically, into spoken or written or imagined utterances and messages. Unless one chooses to believe in some peculiar form of Leibnizian psychophysical parallelism, the immaterial must sooner or later be brought down to the physiological and hence the material level.

Some attempts to circumvent this ineluctable necessity have, to be sure, been made by appealing to the physical uncertainty principle and to forces and phenomena on subatomic levels. These attempts involve misapprehensions and misjudgments which have been well exposed by Rensch (1960) and Nagel (1963), among others. What happens on the subatomic levels may indeed be hard to conceive and to describe in terms habitually used in talking about and in dealing with macroscopic phenomena. There need, however, be nothing inherently acausal about the subatomic phenomena.

The great neurophysiologist Eccles (1953) stressed the extreme smallness of some of the biological structures at which the conjunction of mind and matter may be assumed to take place. These are the synaptic knobs, at which the dendrites (branch fibers) of the different neurons come in contact, and which are estimated to weigh between 10^{-12} and 10^{-13} grams. Yet the excitation across so tiny a mass of matter may result, within a few thousandths of a second, in discharges in hundreds of thousands of neurons. Given so efficient an amplifying device, might we not suppose that an "immaterial" force can exert a powerful influence on macroscopic bodily movements and processes? Here again Heisenberg's uncertainty principle is being invoked. But this is hardly a valid argument; no matter how minute a body a synaptic knob may be, anything that inclines this minute body to act this way or that way must necessarily be another material substance or another physical energy. The fondness felt by some people toward uncertainty principles is something of a misapprehension. We must be looking for directive certainty, not for uncertainty, agencies.

Attempts have also been made to turn the above arguments around. Instead of supposing that immaterial agencies somehow work on matter, assume that all matter has in it some rudiments of life, feeling, and volition. These theories, variously named hylopsychism, panpsychism, panentheism, etc., have been advanced by eminent modern philosophers (Whitehead, Hartshorne), as well as biologists (Rensch, Teilhard de Chardin, Birch, Wright, and others). Rensch's (1960) reasoning runs about as follows. There are no gaps in the evolutionary descent of man and higher animals from the lower ones; there is consequently no ground to deny that all animals, including the unicellular protozoans, have perception of the properties of sensations and have some accompanying feelings; nor can one deny sensations to plants, although "these sensations would not be connected with each other, and therefore would be basic elements essentially different from ours, which are imbedded in a continuous stream of consciousness." The evolutionary continuum extends, however, down to the inorganic realm; hence, molecules and atoms should "also be credited with basic parallel components of some kind." Rejecting all forms of vitalism, Rensch concludes, "Since all phenomena are 'psychic,' there is no contrast of subject and object or of matter and 'soul,' and the natural scientist need not search for an abstract world of reduced matter beyond the realm of general 'consciousness.' "

The argumentation of Wright (1964), although on the whole parallel to that of Rensch (with which Wright is apparently not acquainted), differs in interesting details of emphasis. Wright finds that "The philosophical biologist cannot escape from the problem of the relation of mind and matter," because, "If the nonliving world is completely devoid of mind, and if, as it seems necessary to believe, there was a time when no life could exist, how did mind appear?" Wright is one of the founders of the modern version of the biological theory of evolution, but he nevertheless thinks that "Emergence of mind from no mind at all is sheer magic." To escape this "sheer magic," Wright concludes that "The only satisfactory solution of these dilemmas would seem to be that mind is universal, present not only in all organisms and in their cells but in molecules, atoms, and elementary particles."

Hartshorne (1937, 1962) has generalized panpsychism and panentheism in a philosophical system. All "individual units of nature," i.e., biological individuals, the cells of which they are composed, molecules, atoms, and subatomic particles, are living beings which have some kind of experiencing or feeling. This does not, however, mean that "composites," like rocks, clouds, etc., are also alive; they are composites, not individuals. The seeming inconsistency is resolved by an appeal to the "societal" principle, going back to Whitehead and further to Leibniz—a sentient individual may be an organized society, or a colony, of sentient monads. God is, then, a superindividual, and all other individuals are his constituent parts. Since God is the "cosmic organism," it follows that, "As our enjoyment of health is our participation in the health of the numerous cells, so the happiness of the cosmos is the integration of the lesser happinesses of the parts. The benevolence of God is the only way the psycho-physics of the cosmic organism can be conceived."

We have seen that to Rensch and to Wright the attractiveness of panpsychism lies in that it makes it unnecessary to assume the appearance in evolution of entirely novel qualities. Rensch says "After all, the fundamental problem is whether or not there was a *de novo* formation of the 'psyche.' " The philosophic difficulty involved should not be underestimated, but I believe it should be faced and overcome. The situation has parallels with the explanation of individual development by preformism and epigenesis. The ostensibly simpler view is that the embryo is all there, preformed and hidden in the sex cell, and must merely grow in size to become an adult. On closer examination the simplicity proves deceptive. Although the instructions of the developmental processes are "preformed" in the genes, in the same sense as the instructions for a computer are in the perforations of the computer tape, the development of the embryo is epigenesis, formation of a new body which was not present earlier as an individual.

Viewed in evolutionary perspective, the potentiality of life must have been contained in the inorganic world. The evidence of this is simply that we know that in fact life eventually did appear. Similarly, the potentiality of mind must have been present in the protoplasm, since we know that rational beings did eventually arise. Wright (1964) correctly says, "The fertilized egg, while a living cell, behaves in a way that suggests a purely physico-chemical interpretation rather than one involving mind." But he continues thus: "If the human mind is not to appear by magic, it must be a development from the mind of the egg and back of this, apparently, of the DNA molecules of the egg and sperm nuclei that constitute its heredity." This seems to me a non sequitur. Of course, the potentiality of mind must be present in the egg and the sperm and in the DNA molecules. But it does not follow that eggs and sperms themselves have minds. A stone has in it a potentiality of becoming a statue, but it does not follow that every stone has a statue concealed in it.

It must be granted that the potentiality of a child is, or was, present in his parents. Note, however, that not even the genotype of this particular child was actualized until the two particular sex cells united. With hundreds of egg cells and billions of sperms, the parents contained the potentialities of countless other children as well. Which ones of the countless potentially possible genotypes were to be actualized was not decided (except in the sense of a Laplacian determinism; see below) until fertilization. And even then, an embryo became an infant, and an infant became a person, only by stages, without sharp breaks in the developmental continuity, when a "mind" or a "soul" could be imagined to enter. I see nothing gained by supposing that egg cells and spermatozoa have minds, however attenuated. Nor is there much sense in saying that the packages on grocery store shelves are living and contain a germ of my person, although some of their contents will be so when assimilated in my body.

The assumption that minds have existed and will continue to exist eternally, or at any rate for a very long time, appeals strongly to many people. Annie Besant, the theosophist, was sure that nineteen of her last thirty-six incarnations had taken place in India, and presumably felt assured that some more were to come. Fortunately or unfortunately, such assumptions have little to recommend them by way of validatability or even of tolerable plausibility. At least the outward appearances seem to favor the view that minds arise, develop, and then disappear. The particular mind which is myself arose by degrees, starting some sixty-six years ago, apparently from no mind at all, just from mindless chromosomes, cytoplasm, and nutrient substances. Before then, say seventy years ago, there was no such mind. If this is "sheer magic," it is a kind of magic the world is full of. Within perhaps a few years from now, this particular mind will go out of existence. The alternative is, of course, to follow the example of people who believe that minds do not disappear, but become stored in some places of eternal bliss or of eternal damnation. This is not easy to give credence to; even harder is it to surmise that seventy years ago my mind already existed somewhere offstage, and merely waited for its turn to become implanted in a convenient chromosome or cell.

Hartshorne with a stroke of his pen removes his theory from the realm of science when he writes, "Any unit of nature which acts as one, and is not known to be a mere composite, may, for all anyone could *ever* [Hartshorne's emphasis] know, have its own form of feeling and other psychical traits." An assumption which cannot be falsified by evidence has no standing in science. Although no satisfactory formal definition of life has been proposed, we do know some characteristic phenomena which occur in all living beings. One of them is self-replication (cf. the following chapter). The talk about such "individual units" as atoms or electrons being somehow living may be a poetical metaphor, but it is empty of meaning as a biological hypothesis.

The difficulty with "mind" of a cell or an atom is greater still. The great physiologist Sherrington (1953) states the situation thus:

Though living is analyzable and describable by natural science, that associate of living, thought, escapes and remains refractory to natural science. In fact natural science repudiates it as something outside its ken. A radical distinction has therefore arisen between life and mind. The former is an affair of chemistry and physics; the latter escapes chemistry and physics.

This does not mean that Sherrington was an exponent of some kind of mind vitalism.

The simple fact is that mind and its activity, thought, are experienced, not observed. I am directly aware only of my own mind, and experience only my own thought. I assume that other people also experience something like my thoughts, because their outward behavior shows features analogous to mine, which in myself I know to be accompanied by my thought. But to quote Sherrington again: "We know nothing of any relation between thoughts and the brain, except a gross correlation in time and space." Does this give us license to assume minds and thoughts in molecules and atoms everywhere? We cannot easily prove or definitively refute the assumption that certain animals, say dogs or horses, have some kinds of minds, and in dealing with them we often like to assume that they do. However, the farther away from man we go, the more tenuous the analogy becomes, and correspondingly less satisfactory is the biological and philosophical radical animism.

The evolutionary origin of mind seems to require no "magic." Every one of us knows for certain that his own mind was not always what it is now. Its first flickers arose sometime during childhood; it grew and took its present form by stages and gradations in the process of living. There is nothing that I can see that forbids an analogous process in evolution. The capacity of developing a mind arose in the human species, or in its close or more remote ancestors, and then grew and took form. How gradual this process of growth and formation was is a different question. The evolutionary process does not flow like a quiet river, always at the same speed. It may have periods of relative calm, and others of intense innovation.

The origin of life and the origin of man were evolutionary crises, turning points, actualizations of novel forms of being. These radical innovations can be described as emergences, or transcendences, in the evolutionary process. Neither of these designations is free of undesirable connotations; "transcendence" seems rather preferable, and will be used in the following chapters. Human mind did not arise from some kind of rudimentary "minds" of molecules and atoms. Evolution is not simply an unpacking of what was there in a hidden state from the beginning. It is a source of novelty, of forms of being which did not occur at all in the ancestral states.

Here inevitably arises the thorny question: was it determined that these new forms of being were to appear? And if so, were they real novelties? In his famous statement of mechanistic determinism, Laplace said that,

An intelligence knowing all the forces acting in nature at a given instant, as well as the momentary positions of all things in the universe, would be able to comprehend in one single formula the motions of the largest bodies as well as of the lightest atoms in the world, provided that its intellect were sufficiently powerful to subject all data to analysis; to it nothing would be uncertain, the future as well as the past would be present to its eyes.

To such an intelligence, there can be no novelty in the world, and time is no impediment. Of course, such an intelligence would have to be a divine one. (Hartshorne argued, though, that "not even God can be conceived to know as actual what is merely potential.")

A mere finite and limited human mind may speak of the timeless only in poetic images; Teilhard de Chardin has attempted to speak in such images of the ontological, rather than phenomenological, aspects of evolution. Regardless of whether this attempt was or was not successful, it is something very different from an urge to find gaps in the scientific picture of the world, and particularly of evolution, and to utilize them as evidences of the existence of God. Birch (1965) is undeniably right in saying that "No reconciliation is possible between religious fundamentalism and modern science." Gods of the gaps are dead. And they have been killed by those who most cherished and wanted to protect them. This has been most clearly and explicitly stated by Tillich, a theologian rather than a scientist:

The first step toward nonreligion of the western world was made by religion itself. This was when it defended its great symbols, which were its means of interpreting the world and life, not as symbols, but as literal stories. When it did this it had already lost the battle.

And yet,

What happens in time and space, in the smallest particle of matter as well as in the greatest personality, is significant for the eternal life. And since eternal life is participation in the divine life, every finite happening is significant for God (Tillich 1963).

This is like the saying of an early nineteenth-century theologian, Schleiermacher, that "Miracle is simply the religious name for event. Every event, even the most natural and usual, becomes a miracle, as soon as the religious view of it can be the dominant."

3

Evolution

and Transcendence

The late A. L. Kroeber, for many years the dean of American anthropology, remarked that evolution has now become a "household" word. Kroeber certainly had no illusions that a majority of "householders," even in the most civilized countries, have anything but the vaguest ideas of what evolution really means. Nevertheless, as Sorokin (1937) has pointed out, evolution is "a dominant category of sensate mentality," and

The standpoint of "origin and development and evolution" is our main standpoint in studying anything, from religion to the stock market. It has rooted itself in our mind so deeply that many of us cannot even conceive of any other—nonhistorical, or nonevolutionary, or nondevelopmental—approach to the study of any phenomenon.

To most people the word "evolution" stands for the biological theory propounded by Charles Darwin. In point of fact, biological evolution was discussed by several of Darwin's predecessors, although it was Darwin who finally marshaled an overwhelming body of evidence in support of his theory. Evolution applies, however, to a field far wider than biology. The evolution of the solar system, of the terrestrial globe, and of human societies, these were recognized, or at any rate debated, before biological evolution.

There is no satisfactory general definition of evolution. "Sustained change" comes probably as close as possible at present. In the special case of biological evolution this may be amended to become "sustained change over a succession of generations," to differentiate the evolutionary development (phylogeny) from the development of an individual (ontogeny).[1]

A thoroughgoing denial of all evolution is hardly possible, although Parmenides believed that all change was illusion of the senses. However, even if there existed some supernal changeless reality, it might be asked how extensive were the changes which our illusions have made us believe happened. Almost all people, down to the most primitive, have composed creation myths to account for the origins of the world and of mankind. These myths are often invested with exquisite poetic feelings, and according to some theories they stem from the greatest depths of the human psyche (Long 1963). The creation myths narrate mostly the deeds of divine personages, which resulted in the origination of the world of today. The heroic deeds and the turbulent events came to pass, however, mainly at the world's beginnings. Once set in operation, the world is usually pictured as a relatively stable and immutable place.

If taken by itself, the account of the creation given in the first chapter of the Book of Genesis can be interpreted, and it was in fact so interpreted, as a basis for an unhistorical and nonevolutionary conception of the nature of

1 Huxley (1955) proposed a much more elaborate, and he believed general, definition of evolution: "Evolution is a self-maintaining, self-transforming, and self-transcending process, directional in time and therefore irreversible, which in its course generates ever fresh novelty, greater variety, more complex organization, higher level of awareness, and increasingly conscious mental activity." This, unfortunately, will not do even for biological evolution alone. Instances of evolutionary stagnation show that evolution is not self-maintaining; evolution is primarily a groping and only secondarily a directional process; evolution is not infrequently regressive rather than progressive; cladogenesis (increasing variety) is as frequent as anagenesis (change within a single evolutionary unit); consciousness or self-awareness have arisen, as far as we know, in only a single line, that leading to man.

the universe and of man. This unhistorical conception prevailed until it was shattered by the impact of scientific discovery in post-Renaissance times. Not all change was denied, but the recency of the assumed date of creation, as in Bishop Ussher's chronology, naturally implied that the world could not have changed very greatly since it came into existence. Hence, in the words of Ecclesiastes: "The thing that hath been, it is that which shall be; and that which is done is that which shall be done; and there is no new thing under the sun."

Another class of accounts of creation assume eternal recurrence of creations and dissolutions of the universe. In a sense, this is combining evolutionist and nonevolutionist views. The cosmology of the Vishnu Purana is an excellent example of this species. The universe exactly repeats itself in cycles of 311,040 billion years—a colossal figure even compared to the modern estimate of the age of the universe, a mere 5 billion years! Within this grand cycle, there are shorter subordinate cycles of 4,320,000 years; they nevertheless involve radical changes, from a creative beginning (Krita Yuga) to a destructive close (the era of Kali, in the midst of which we find ourselves at present). Cyclic theories were also in vogue with the Pythagoreans; the length of the cycle (*annus magnus*) was variously estimated from 5552 to 100,020 years. In modern times, Nietzsche was the most passionate exponent of the belief in eternal return; Danilevsky, Spengler, and to a certain extent Sorokin favored cyclic theories of history.

Christianity is, among the great religions, most explicitly history-conscious, and in this sense evolutionistic. It affirms that the history of mankind and of the world is not merely an illusion or an irremediable evil. History is the vehicle of creation. The world had a beginning, and will have an end. At a certain point in history an event of pivotal significance took place—God assumed human form and lived among men. Man stands in the center of history's meaning. In Tillich's (1963) words, "In the Christian vocational consciousness, history is affirmed in such a way that the problems implied in the ambiguities of life under the dimension of history are answered through the symbol 'Kingdom of God.' " This did not necessarily imply a doctrine of progressive evolution toward a perfect state; in fact, Christianity has inherited from Judaism the tenet of an initial perfection followed by a catastrophic fall.

It is therefore not an accident that the idea of progress grew and developed on the Judeo-Christian cultural background, although largely in secular rather than in religious contexts (Bury 1932). The Encyclopedists generally believed in "the total mass of the human race moving always slowly forward." Voltaire thought that man advances "from the barbarous rusticity" to "the politeness of our era." The Age of Enlightenment closes with the first full-fledged theory of progressive evolution of mankind, that of Condorcet. This is a hymn of unbounded optimism composed in the shadow of death. On a different philosophical basis, the evolutionary ideas of Herder, Kant, and Fichte lead to Hegel's system, in which history is understood as progressive manifestation of the Spirit. Marx puts Production and Economics in the place of the Spirit, and comes up with substitutes for the Kingdom of God, called Socialism and Communism. Marx discovered his "evolution" of society at about the same time as Darwin published his account of biological evolution. Marx had the perspicacity to recognize in Darwin a fellow evolutionist, an affiliation which Darwin was most reluctant to accept. Soon thereafter, Spencer, Tylor, Morgan, and others were founding evolutionary social anthropology, expressly built on Darwinian theoretical premises.

Theories of cosmic and terrestrial inorganic evolution also appeared well ahead of organic evolution. Before he became the prince of the philosophers, Kant at a relatively youthful age advanced the primal nebula theory (1755). Newtonian gravitational forces had, Kant thought, gradually assembled the originally scattered matter into sun and

planets. Laplace in 1796 and 1808 envisaged the primordial
sun as having been a rapidly rotating sphere of gas. Parts
of this sphere were torn off by powerful centrifugal forces,
and condensed into planets and moons. The doctrine of
uniformitarianism, put forward by Buffon and by Hutton
in the eighteenth century, was developed by Lyell in the
nineteenth as the basis for the study of the geological
changes of the earth. Skeptical at first of biological evolution, Lyell inspired and supported Darwin and Darwin's
theory.

Modern cosmological theories are, however, quite recent.
They are products of the twentieth-century advances in
physics and astronomy. They fall into two major groups
which, at least to an outsider, seem to be very different
and incompatible. A final decision between them appears
not yet possible, making this an exciting time in cosmology.
Steady-state theories have been developed by Bondi, Gold,
and especially by Hoyle. Their basic assumption is that new
matter continuously arises in the expanding universe. The
rates of expansion and of creation of new matter are correlated, making the density of matter in space approximately constant in time. New galaxies are continuously
being formed. The other theories assume that the universe
began 5 to 10 billion years ago from a state of extreme
compression, when the density of its matter and radiation
was some 100 million million times the density of water.
The beginning was a stupendous explosion (hence the
nickname "The Big Bang" theory, associated particularly
with the theory proposed by Gamow). Within five minutes,
when the temperature was of the order of a billion degrees,
atoms of chemical elements began to be formed, the relative abundance of the elements at present being the
product of that process, estimated to have taken only about
twenty-five minutes. The steady-state theory assumes, on
the contrary, that the elements heavier than hydrogen and
helium are constantly being "cooked" in the interior of
very hot stars.

The evolutionary origin of matter may, thus, be something which occurred in a far-distant past, or a process still operating under certain special conditions at present. On the other hand, while stars were regarded as symbols of permanence for time without end, it seems now to be established beyond reasonable doubt that stars undergo evolutionary changes. Furthermore, at least if present theories are to be trusted, these changes mostly follow a fairly predictable path ("The main-sequence," Hoyle 1955). Our sun is expected to increase greatly in brightness, because of an expansion in volume connected, however, with a decrease of the surface temperature. At its maximum, the sun will be 200 to 300 times its present size. This will be followed by a shrinkage, but not below the present size, and a great increase of temperature. The nuclear "fuel" will then approach the depletion stage and the production of energy will eventually cease.

Biological evolution is the middle term of the evolutionary triad—cosmic, biological, and human. Reference has already been made above to the fact that biological evolution was discovered after the other two. Darwin was the real discoverer, although he had several important predecessors, of whom Lamarck was the most prominent, though equivocal. To locate and to study these anticipations is a fascinating task, but it is losing perspective to say, as some recent writers have done, that Darwin merely sorted out ideas borrowed from others. Moreover, Darwin's theory has stood the test of time remarkably well. Modern biological theory is, of course, far from being identical with Darwin's of more than a century ago, and yet there is an unbroken intellectual continuity between them. The modern theory is called "biological," or "synthetic," because it represents a convergence and synthesis of findings and interpretations of almost all the biological disciplines. This is in itself a novelty, because until about thirty years ago geneticists, systematists, paleontologists, embryologists, and others were expounding theories which seemed, to them, to fit their particular materials, but took scant notice of the neighboring disciplines.

The essence of the biological theory is fairly simple, and it will be discussed in a different context in Chapter 6. No more than a very brief sketch is needed here. The fundamental postulate is that evolution consists mainly of responses of a biological species to the challenges of its environments. Life is, in general, an improbable state of matter; it is really a tour de force achieved against heavy odds, by means of a slow ascent through evolution. If it is not to be snuffed out by hostile environments, life must at all times maintain, and whenever possible improve, its adaptedness to its surroundings. Changing environments present the severest challenges, since it is quite unlikely that the genetic endowments formed in response to the old environments will be, by accident, fully suited to the new ones.

The method by which a living species responds to environmental challenges is natural selection. This was clearly envisaged by Darwin, but the present understanding of natural selection is a product of newer biology. The raw materials of evolution are the genetic variants which arise by mutation. The astounding property of the mutation process is its adaptive ambiguity. Mutants appear regardless of whether they are or can ever be useful, instead of only suitable mutants arising where and when needed. In fact, most of them are harmful. Mutation alone, uncontrolled by natural selection, could only result in degeneration, decay, and extinction.

Natural selection has often been compared to a sieve, which retains the few useful mutants and lets the harmful ones be lost. This analogy is seriously misleading. It leaves out of account sex and gene interaction. Sexual reproduction generates ever-new combinations of the genetic variants which appeared originally as mutations. This is far more important than mere shuffling and reshuffling of independent units would be; the usefulness or harmfulness of a genetic variant often depends on the genetic system in which the variant is placed. A variant which is beneficial in some combinations may be detrimental in others.

To make the sieve analogy valid, one would have to imagine an extraordinary "sieve." It must be contrived to retain or to discard particles not on account of their size alone, but in consideration of the qualities of all other particles present. We have, then, not a sieve but a cybernetic device which transfers to the living species "information" about the state of its environments. This device also makes the evolutionary changes that follow dependent upon those which preceded them. The genetic endowment of a living species contains, therefore, a record of its past environments, as well as an imprint of the present one. This genetic endowment is not a mosaic of genes with autonomous effects; it is an integrated system, the parts of which must fit together to be fit to survive.

The idea that man and the world which he inhabits are products of an evolutionary development seems, strangely enough, "degrading" to some people. It is, on the contrary, a prerequisite of humanism in Tillich's sense (page 4), since to have potentialities to be actualized man must first of all have the potentiality of evolving. He must be, individually and collectively, not a state but a process. The cosmos must be the cosmogenesis. And, in the words of Teilhard de Chardin (1964), "We see more clearly with every increase in our knowledge that we are, all of us, participants in a process, Cosmogenesis culminating in Anthropogenesis, upon which our ultimate fulfillment—one might even say, our beatification—somehow depends."

The question which presents itself is whether the cosmic, the biological, and the human evolutions are three unrelated processes, or are parts, perhaps chapters or stages, of a single universal evolution. Asking this is really another way of posing the questions, how living bodies arose from the nonliving matrix, and how the stream of consciousness and self-awareness started from the straightforward physiological processes in our animal ancestors. How difficult these questions are can be seen from the following statement by an eminent biologist:

What is still so completely mysterious as to acquire for many human beings a mystical quality, is that life should have emerged from matter, and that mankind should have ever started on the road which so clearly is taking it farther and farther away from its brutish origins (Dubos 1962).

Here it is necessary to guard against two oversimplifications, opposite in sign but equally misleading. One assumes complete breaks in the evolutionary continuity between life and nonlife, and between humanity and animality. The other overlooks the differences between the cosmic, biological, and human evolutions, and thus loses sight of the origin of novelty. The best hope of making the problem manageable lies, it seems to me, in using the concept of levels, or dimensions of existence, developed by dialectical Marxists on the one side and by the great theologian Paul Tillich on the other.

Stated most simply, the phenomena of the inorganic, organic, and human levels are subject to different laws peculiar to those levels. It is unnecessary to assume any intrinsic irreducibility of these laws, but unprofitable to describe the phenomena of an overlying level in terms of those of the underlying ones. One of the Soviet high priests of Marxism, Present (1963), expresses this fairly clearly as follows:

Wherever it arose, human society must have come from the zoological world, and it was work, the process of production, that made man human. However, what removed people from the animal way of life and gave a specificity to their [new] life, became the essence and the basis of the history that ensued. . . . Likewise, in the realm of living nature, what removed the novel form of the material motion from its nonliving prehistory necessarily became its essence, its fundamental basis.

According to Tillich (1963),

No actualization of the organic dimension is possible without actualization of the inorganic, and the dimension of the spirit would remain potential without the actualization of the inorganic. . . . All of them are actual in man as we know him, but the special character of this realm is determined by the dimensions of the spiritual and historical . . . the dimension of the organic is essentially present in the inorganic; its actual appearance is dependent on conditions the description of which is the task of biology and biochemistry.

Inorganic, organic, and human evolutions occur in different dimensions, or on different levels, of the evolutionary development of the universe. The dimensions or levels are, to be sure, not entirely sundered from one another; on the contrary, there are feedback relationships between the animate and the inanimate, and between the biological and the human. Nevertheless, the different dimensions are characterized by different laws and regularities, which are best understood and investigated in terms of the dimension to which each belongs. The changes in the organic evolution are more rapid than in the inorganic. Inorganic evolution did not, however, come to a halt with the appearance of life; organic evolution is superimposed on the inorganic. Biological evolution of mankind is slower than cultural evolution; nevertheless, biological changes did not cease when culture emerged; cultural evolution is superimposed on the biological and the inorganic. As stated above, the evolutionary changes in the different dimensions are connected by feedback relationships.

The attainment of a new level of dimension is, however, a critical event in evolutionary history. I propose to call it evolutionary transcendence. The word "transcendence" is obviously not used here in the sense of philosophical transcendentalism; to transcend is to go beyond the limits of, or to surpass the ordinary, accustomed, previously utilized or well-trodden possibilities of a system. It is in this sense that Hallowell (1961) wrote, "The psychological basis of culture lies not only in a capacity for highly complex forms of learning but in a capacity for transcending what is learned, a potentiality for innovation, creativity, reorganization and change." Erich Fromm (1959) wrote that man "is driven by the urge to transcend the role of the creature," and that "he transcends the separateness of his individual existence by becoming part of somebody or something bigger than himself."

Cosmic evolution transcended itself when it produced life. Though the physical and chemical processes which occur in living bodies are not fundamentally different from those found in organic nature, the patterns of these processes are different in the organic and inorganic nature. Inorganic evolution went beyond the bounds of the previous physical and chemical patternings when it gave rise to life. In the same sense, biological evolution transcended itself when it gave rise to man. There obviously exist phenomena and processes, ranging from self-awareness to the human forms of society and of history, which occur exclusively, or almost exclusively, on the human level. It seems unnecessary to labor the point that a great range of potentialities are open to man only.

The origin of life and the origin of man are, understandably, among the most challenging and also most difficult problems of evolutionary history. It would be most unwise to give a fictitious appearance of simplicity to these singularly complex issues. On the other hand, so much has been learned about them, not only in the last fifty but even in the last ten years, that a bird's-eye view of the present status may be useful (for more detailed accounts, see Calvins 1964, Fox 1965, Oparin 1961, Urey 1952, and especially the very readable summary in the book of Sullivan 1964). Attempts to find extraterrestrial life are perhaps not far from realization, and conceivably from success. Regardless of one's evaluation of the chances of such success, one must agree with Lederberg (1965) that "By every standard, this is an epochal enterprise: a unique event in the history of the solar system and of the human species, and the focus of an enormous dedication of cost and effort."

Life on earth is reckoned to be two billion or more years old. It first arose under environmental conditions quite different from those which exist today; these conditions can be reconstructed only with difficulty. As the earth cooled sufficiently for the water vapor to condense and to form the oceans, the atmosphere consisted of such gases as hydrogen, methane, ammonia, carbon dioxide, with lesser quantities of nitrogen and little free oxygen. Chemical reactions that can take place in such mixtures have been extensively studied in laboratories. Formaldehyde, acetic acid, succinic acid, and as many as ten different amino acids can all be formed. These, and other even more complex substances, now made chiefly or only in living bodies, could arise and accumulate in solution in the then lifeless oceans, making the latter a kind of dilute "soup" of organic compounds. According to Ponnamperuma (in Fox 1965), these might have included the purines adenine and guanine, the sugars ribose and deoxyribose, and the nucleoside adenosine. These particular compounds are especially interesting, because they are among the constituents of the nucleic acids, DNA and RNA.

Haldane (in Fox 1965) describes life as "indefinite replication of patterns of large molecules." Much ingenuity has been used to construct models of processes that might conceivably have led to combining the smaller molecular constituents into large molecules, such as DNA, which can facilitate the synthesis of their copies. The best that one can say about these models is that some of the processes which they postulate might conceivably have happened, but not that they did actually happen in the real history of the earth. It is also not quite generally agreed that self-reproduction is both the necessary and sufficient condition for regarding a chemical system as living, though it is agreed that it is at least a necessary condition. The reason for this is as follows: Self-replication means not only that a system engulfs suitable materials from its environment (food) and transforms them into its own likeness (reproduction), but also that any changes that may occur in the self-replicating process (mutation) may be reflected in the relative frequencies of the changed and unchanged systems in the course of time (natural selection). In other words, the origin of self-replication opens up the possibility of biological evolution. This evolution may, although it need not necessarily, be progressive. The evolution of a self-replicating system may be cut short by exhaustion of the environment or by changes in the system which make it less efficient (extinction). It is possible that, if self-replicating systems arose repeatedly, most of them were lost without issue. The point is, however, that at least one system was preserved and "inherited the earth," by becoming the starting point of biological evolution.

The evolutionary transcendence which rose from the inorganic level or dimension to the organic may thus conceivably be explained in scientific, more precisely in biological and biochemical, terms. This does not mean that it has already been so explained. As Warren Weaver (1964) said, "A person usually considers a statement as having been explained if, after the explanation, he feels intellectually comfortable about it." The existing accounts of the origin of life leave one uncomfortable, and this not so much because the information available is incomplete, but rather because they involve a curious form of the logical error known as *petitio principii*. On the biological level, the process of natural selection is said to bring about the realization of what would be utterly improbable without it. As pointed out in the foregoing chapter, man consists of molecules and atoms, of transformed "groceries," and in the individual development of a man, these "groceries" do come together in just the right ways to compose a man. They do so because an individual is a link in a chain of generations extending back to the origin of life, and the natural selection has stored in these generations the instructions needed for the individual development to take place.

If natural selection has achieved this feat in biological evolution, it becomes hard to resist the temptation to invoke the help of natural selection to explain the origin of biological evolution as well. Was there some prebiological form of natural selection to contrive the simplest living systems from the chemical components which had already achieved a certain degree of complexity in the diluted "soup" of the ancient oceans? Oparin, Bernal, Fox, and others have devised a number of imaginative schemes of "the beginning of the elements of selection," which may have been the "starting point on the basis of which Darwinian selection was arising." Pattee (in Fox 1965) speaks of chemical evolution "which implies only stochastic reactions without hereditary order or replication," of heredity as "any process by which information is transmitted from one structure (the parent) to another structure (the heir) so as to result in a net increase in the physical order of the total system," and of evolution as "the propagation and gradual increase of hereditary information in the course of time." Having thus redefined the concepts to suit his convenience, Pattee comes to the conclusion that "it is not productive to ask when hereditary propagation of order became replication or self-replication, or when direct feedback interactions with the environment became natural selection. The difference is only in degree, and both evolved continuously." Unfortunately, this is begging the question; differences in degree grow large enough to become differences in kind. How and when natural selection arose is precisely the crux of the problem of the origin of life. Natural selection is differential reproduction; when two forms of life reproduce at ratios different from unity natural selection is operating, but for natural selection to operate there must be reproduction, and reproduction is the key property of life.

In these days of space exploration, the problem of the origin of life has rather suddenly become fashionable. Belief in the existence of extraterrestrial life, and of extra-terrestrial rational beings, has a potent romantic appeal. Books have been written, and at least one conference held, to discuss in all seriousness how best to catch the signals which the denizens of cosmic space may be beaming at us, and how best to go about trying to communicate with them. It is almost a pity to oppugn so glamorous a notion. But what are its credentials? Probably the main one is that there must be, conservatively estimated, 100 million planets in the universe whose environments are believed by authoritative astronomers to be tolerably similar to terrestrial ones. And with so many chances available, even an improbable event is likely to happen more than once. This would be a strong argument if we knew just how probable, or improbable, the event under consideration really is, but for this we have no secure enough basis for judgment.

The further argument, that once life has arisen it is bound to evolve approximately as it did on earth is less dependable still. At first blush, this may not seem unsound; many of the body structures of existing organisms are adaptations which help to solve the problems of survival posed by the environment, and it may seem that natural selection should bring them into existence wherever they are needed, as it did on earth. This is not necessarily so; in the first place, an adaptive problem may not be solved at all, and the species may become extinct; moreover, and this is crucial, many adaptive problems may be solved in more than one way. The solutions which we know to have been utilized by organisms on earth cannot be guaranteed to have been either the best conceivable ones, or the ones that would occur again in the thinkable (not, however, probable) situation of the evolution occurring for the second time here, on earth. (For further arguments concerning the unlikelihood of extraterrestrial "humanoid" life, see Dobzhansky 1960 and Simpson 1964.)

I do not wish to be understood as maintaining that the origin of life involved agents or processes that did not also earlier or later operate on earth. The flow of evolutionary events is, however, not always smooth and uniform; it also contains crises and turning points which, viewed in retrospect, may appear to be breaks of the continuity. The origin of life was one such crisis, radical enough to deserve the name of transcendence. The origin of man was another. This should not be taken to mean that the origin of life or of man was instantaneous or even very swift. A process which is very rapid in a geological (more precisely, paleontological) sense may appear to be lengthy and slow in terms of a human lifetime or a generation.

The appearance of life and of man were the two fateful transcendences which marked the beginnings of new evolutionary eras. They were, however, only extreme cases of radical innovations, other examples of which are also known. The origin of terrestrial vertebrates from fishlike ancestors opened up a new realm of adaptive radiations in the terrestrial environments, which was closed to water-dwelling creatures. The result was what Simpson (1953) has called "quantum evolution," an abrupt change in the ways of life as well as in the body structures. Domestication of fire and the invention of agriculture were among the momentous events which opened new paths for human evolution. In a still more limited compass, the highest fulfillment of an individual human life is self-transcendence.

While the origin of life on earth is an event of a very remote past, man is a relative newcomer (even though the estimates of his antiquity have been almost doubled by recent discoveries). The record is still fragmentary, but the general outlines of at least the outward aspects of man's emergence are recognizable. During the early part of the Pleistocene age (the Villafranchian time, perhaps two million to one million years ago), there lived in the east-central and in the southern parts of the African continent at least two species of *Australopithecus,* a genus of Hominidae, the family of man. One of these, larger in body size (*Australopithecus robustus,* and its race *boisei*), apparently represented an evolutionary blind alley, and eventually died out. The other species, of a more supple build (*Australopithecus africanus*), may have been one of our ancestors.

There is good evidence in the structure of their pelvic bones that both species walked erect; their teeth were, if anything, more like ours than like those of the now-living anthropoid apes. The brain case capacity, though large in relation to the body size, was well within the ape range. Perhaps both species of *Australopithecus,* or at any rate one of them, made and used primitive stone tools. The remains of a most interesting creature have recently been found in the Villafranchian deposits in east-central Africa, and given the name *Homo habilis.* The name connotes the opinion of its discoverers that this creature had already passed from the genus *Australopithecus* to the genus *Homo,* to which we also belong. On the other hand, it is closely related to *Australopithecus africanus,* and may even have been only a race of that species. Regardless of the name by which it is classified, this is rather clearly one of the "missing links," which are no longer missing.

Later, during the mid-Pleistocene, there lived several races of the species *Homo erectus,* clearly ancestral to the modern *Homo sapiens.* Remains of *Homo erectus* have been found in Java, in China, in Africa, and probably also in Europe. Roughly 100,000 years ago, during the last, Würm-Wisconsin, glaciation, the territory extending from western Europe to Turkestan and to Iraq and Palestine was inhabited by variants of the Neanderthal race of *Homo sapiens.*

Rough stone tools have been found in association with australopithecine remains both in east-central and South Africa. *Homo erectus* in China is the oldest known user of fire. The Neanderthalians buried their dead. These are evidences of humanization. All animals die, but man alone knows that he will die. As will be discussed in more detail in the following chapter, a burial is a sign of a death awareness, and probably of the existence of ultimate concern. The ancestors of man had begun to transcend their animality perhaps as long as 1,700,000 years ago. The process is under way in ourselves. Nobody has characterized this process more clearly than Bidney (1953):

Man is a self-reflecting animal in that he alone has the ability to objectify himself, to stand apart from himself, as it were, and to consider the kind of being he is and what it is that he wants to do and to become. Other animals may be conscious of their affects and the objects perceived; man alone is capable of reflection, of self-consciousness, of thinking of himself as an object.

And according to Hallowell (1960):

The great novelty, then, in the behavioral evolution of the primates, was not simply the development of a cultural mode of adaptation as such. It was, rather, the psychological restructuralization that not only made this new mode of existence possible but provided the psychological basis for cultural readaptation and change.

In 1871, Darwin wrote: "The difference in mind between man and the higher animals, great as it is, certainly is one of degree and not of kind." This is about the state of the issue at present. The estimates of the importance of this difference have, however, varied greatly; it is belittled by those intent on proving that man is nothing but an animal, and overdrawn by those who place him outside the order of nature. To the former, there is no great difference between fish and philosopher, and nothing more remarkable about his human brain secreting thoughts than about his liver secreting bile. Leslie White, an anthropologist, believes that

whether a man—an average man, typical of his group—believes in Christ or Buddha, Genesis or Geology, Determinism or Free Will, is not a matter of his own choosing. His philosophy is merely the response of his neuro-sensory-muscular-glandular system to the streams of cultural stimuli impinging upon him from the outside.

However, the same author in the same book (White 1949) also says,

Because *human* behavior is symbol behavior and since the behavior of infra-human species is nonsymbolic, it follows that we can learn nothing about human behavior from observations upon or experiments with the lower animals.

A. R. Wallace, the codiscoverer with Darwin of the evolutionary role of natural selection, thought that man's mind must have been implanted by a supernatural agency. Lack (1957), one of the foremost modern evolutionary ecologists, is of the same opinion: "A Christian, agreeing to man's evolution by natural selection, has to add that man has spiritual attributes, good and evil, that are not a result of this evolution, but are of supernatural origin." Brunner (1952) may be taken as representative of a type of thinking among modern theologians who concede that man is a product of biological evolution, but nevertheless insist that he has a property called the *humanum,* which is solely his attribute. The *"humanum* is characterized by something which is entirely lacking in the animal, subjectively speaking by the spirit [Geist], and objectively by the creation of culture. . . . It possesses a dimension which is lacking in biology, the law of norms, the faculty of grasping meaning, freedom, responsibility" (a critique of Brunner's views in Birch 1957 and 1965).

Lists of distinctive anatomical, physiological, and psychological characteristics of the human species have often been compiled. Man is an erect—walking—primate, his brain is large in relation to the body size, his hands are fit for tool manipulation, toolmaking, and for carrying objects. He engages in play, is capable of abstract thought, laughter, formation and use of symbols, of learning and using symbolic language, of learning to distinguish between good and evil, to feel reverence and piety. At least some of these characteristics belong to the *humanum*. How did they arise in evolution? It is most certainly difficult, perhaps quite impossible, to reconstruct, even if we were given more detailed data than are now available, the exact sequence of events during the critical stages of the emergence of mankind from prehuman ancestors. There are, however, some comprehensive biological regularities which make a general understanding of the main trends of the evolutionary history of man not out of reach.

If a descended species is found to differ from its ancestor in several characteristics, the origin of each of these characteristics is not necessarily independent. The characteristics may vary together, for at least two reasons. They may be manifold effects of the same genetic factors (as are the different parts of the syndromes in many hereditary diseases and malformations dependent on single genes). Or else, they may be functionally related. There was some discussion among anthropologists as to whether the development of an erect posture and manual skill preceded or followed tool use and toolmaking. A similar issue can be raised concerning the capacity for abstract thinking, formation of symbols, symbolic language, and the beginnings of cultural transmission. To a considerable extent, if not at all points, such issues are spurious. The product grows with the instrument, and the instrument with the product. Hands freed from walking duties by an erect posture can more easily develop the dexterity needed for the manipulation of tools. An erect posture does not, however, guarantee that the anterior extremities will use tools; for example, the giant kangaroos of Australia do not utilize their front paws either for running or for handling tools. Conversely, some monkeys and apes have fairly versatile hands but do not walk upright. It is reasonably clear that an animal which begins to handle tools will derive an advantage from having a pair of its extremities become adept at such operations;

vice versa, an animal which becomes bipedal may profit by using the second pair of its extremities for something else. The crux of the matter is, however, this: increasing tool use puts greater and greater selective premium on bipedalism, and vice versa.

Human languages are very different forms of behavior from the so-called animal "languages," although both serve, of course, the function of communication. This complex, and to many people puzzling, issue, has been ably clarified by Hockett (1959) and Hockett and Ascher (1964). Very briefly, the differences are as follows. Animal calls or signals are mutually exclusive, meaning that the animal may respond to a situation by one or another of its repertory of calls, or may remain silent. Language is productive, i.e., man emits utterances never made before by himself or by anyone else, which are nevertheless understandable to speakers of the same language. People can speak of things that are out of sight, and of past, future, and imaginary things. This is called the property of displacement. Moreover, language has the property of a duality of patterning. It consists of units, phonemes, which have no meaning in themselves but serve to compose utterances that are meaningful. The capacity to learn a language is biologically inherited; it is a capacity to learn (at least in childhood) any one of the human languages. A chimpanzee cannot do it, although his larynx is apparently capable of producing all the necessary sounds; the incapacity is caused by the absence of certain brain mechanisms. Human languages consist of utterances the meaning of which is socially established by convention. Neologisms and changed habits of speech constantly appear.

Animal signals, calls, and the "dances" of the honey bees are admirably efficient methods of communication, but only within the narrow compass of the needs of the particular species in which they occur. There seems little point in belaboring the truism that human speech possesses the advantage of an almost infinitely greater versatility. This is because man's is a symbolic language. Cassirer (1944), Langer (1948) and some others, consider the ability to form and to use symbols man's most important distinctive quality. In Cassirer's words "Signals and symbols belong to two different universes of discourse: a signal is a part of the physical world of being; a symbol is a part of the human world of meaning." And further:

The principle of symbolism with its universality, validity, and general applicability, is the magic word, the Open Sesame giving access to the specifically human world, to the world of human culture. Once man is in possession of this magic key further progress is assured.

The capacity to form and to operate with abstract ideas and symbols is correlated in evolution, if not in physiology, with the capacity to use human language. Here too, the product grows with the instrument, and vice versa. And these capacities are, in turn, correlated with toolmaking. It should be noted that tool using and toolmaking are performances almost as profoundly different as signs and symbols. To make a tool for a future employment one needs more than manual dexterity; what is necessary is formation of a mental picture of a situation which is expected to arise in the future but which is not yet given to the senses. In evolution, all these capacities were connected by feedback relationships. An advance in any one of them made the others adaptively more valuable, and advances in these others conferred higher selective premiums on further developments of the first. Owing to these cybernetic interrelations, the protohuman species, gradually shifting from animal to human ways of life, eventually reached a point of no return. To revert from the incipient humanity back to animality would have been genetically difficult, even if this were adaptively profitable. Extinction would become more likely than an evolutionary retreat.

At this point it is necessary to be reminded of some biological evolutionary principles which were mentioned above in another connection. Evolution arises through the action of natural selection in response to the exigencies of the environment. It is a utilitarian process; it maintains or enhances the adaptedness to the environment. But natural selection is not some sort of benevolent ghost; it is automatic, blind, and lacking foresight. It is opportunistic, in the sense that it adapts the organism to the environments existing at the time it acts, and it cannot take into account any possible changes of the conditions in the future. As pointed out above, the consequence of this opportunism and myopia may be extinction. The adaptation is liable to drive the species into a blind alley and to make retraction impossible (see further in Chapter 6).

Another consequence of opportunism is even more relevant in human evolution. The fitness that is selected is the overall fitness of the organism to survive and to reproduce, not the excellence of different organs, processes, and abilities taken separately. A consequence is that, especially in radical evolutionary reconstructions, the emerging product is an appalling mixture of excellence and weakness. That this is the case with man is almost a platitude. To quote Hockett and Ascher (1964):

Language and culture, as we have seen, selected for bigger brains. Bigger brains mean bigger heads. Bigger heads mean greater difficulty in parturition. Even today, the head is the chief troublemaker in childbirth. The difficulty can be combatted to some extent by expelling the fetus relatively earlier in its development. There was therefore a selection for such earlier expulsion. But this, in turn, makes for a longer period of helpless infancy—which is, at the same time, a period of maximum plasticity, during which the child can acquire the complex extra-genetic heritage of its community. The helplessness of infants demands longer and more elaborate child care, and it becomes highly convenient for the adult males to help the mothers. Some of the skills that the young males must learn can only be learned from the adult males. All this makes for the domestication of fathers.

The structure of the human psyche discovered by Freud and his followers is a bundle of contradictions and disharmonies, which at first sight seems anything but a product of adaptive evolution. It must, however, be viewed not by itself but within the total evolutionary context. So considered, man is not merely a successful biological species but the most successful one that biological evolution has ever produced. To quote Hockett and Ascher again:

As soon as the hominids had achieved upright posture, bipedal gait, the use of hands for manipulating, for carrying, and for manufacturing generalized tools, and language, they had become men. The human revolution was over.

The biological evolution has transcended itself in the human "revolution." A new level or dimension has been reached. The light of the human spirit has begun to shine. The *humanum* is born.

It remains to consider briefly some of the misgivings which arise in connection with the above account of the evolutionary transcendence giving rise to man. Those who see an unbridgeable gap between the *humanum* and the prehuman state question the presence on the animal level even of rudiments from which the *humanum* could arise. Now, the point which the believers in unbridgeable gaps miss is that the qualitative novelty of the human estate is the novelty of a pattern, not of its components. The transcendence does not mean that a new force or energy has arrived from nowhere; it does mean that a new form of unity has come into existence. At all events, no component of the *humanum* can any longer be denied to animals, although the human constellation of these components certainly can. In recent years this problem has been studied by means of many brilliantly conceived and executed experiments (reviews in Rensch 1960, Thorpe 1962, 1963). Birds and mammals are demonstrably able to form "abstract averbal concepts," such as those of number. In the experiments of Koehler, a bird (raven) learned to choose from among five dishes covered with cardboard disks marked with two to six spots, the dish which contained food. After training, the bird was able to pick the right dish when shown a signboard with the corresponding number of dots. Another bird (a parrot) selected the right dish when given a corresponding number of acoustic stimuli.

The "dance language" of bees is symbolic, indicating the direction of a food source by the movements of the informant bee on the honeycomb. It does not, of course, have any other of the properties of human languages discussed above. "Insight learning" has been demonstrated in chimpanzees by Koehler, Ladygina-Kots, and others. Thorpe writes that

with birds as with men, the perceptions of the experienced and mature individual are built up by a process of perceptual learning whereby the primary conceptions are combined and built into more complex gestalts; indeed, no essential differences between the principles underlying the two processes can be detected.

True play not only occurs in animals as well as in humans, but it performs the important function of training the young in preparation for adult activities. "Protoesthetic" impulses have been shown to exist in monkeys and apes, which can be made to engage in painting, with results not too dissimilar from some products of avant-garde human painters. Some precursors of cooperative, ethical, and even altruistic behavior have repeatedly been alleged to exist in animals, but in the nature of the case they are hard to prove beyond doubt (see Thorpe, *loc. cit.,* for critical reviews).

Another difficulty with the evolutionary explanation of the origin of man is of a more general sort. The occurrence of mutation and of natural selection is said to be due to "chance." Can one possibly believe that the *humanum* arose through summation of a series of accidents? De

Cayeux (1958), a French paleontologist, emphatically disagrees: "The lottery does not suffice. Mutationism is a double failure. The evolution of species cannot be due to chance alone, even if corrected by the selection." Protestations to the same effect have been made with considerable eloquence by such writers as Joseph Wood Krutch, Barzun, and others. Now, a goodly portion of all this is sheer misunderstanding. What, indeed, is meant by "chance," or "randomness," in evolution? Reference has already been made above to the fact that mutations are adaptively ambiguous. Nature has not seen fit to make mutations arise where needed, when needed, and only the kind that is needed. A geneticist would be hard put to envisage a way to accomplish this. But not even mutations are random changes, because what mutations can take place in a given gene is evidently decreed by the structure of that gene. Mutations do not, however, constitute evolution; they are only raw materials for evolution. They are manipulated by natural selection.

Natural selection is a chance process (despite the misplaced superlative "fittest" in the "survival of the fittest") only in the sense that most genotypes have not absolute but only relative advantages or disadvantages compared to others. Natural selection may act if one genotype leaves 100 surviving offspring compared to 99 left by the carriers of another genotype. Otherwise natural selection is an antichance agency. It makes adaptive sense out of the relative chaos of the countless combinations of mutant genes. And it does so without having a will, intention, or foresight. The classical analogy between the action of natural selection and that of a sieve is, as pointed out above, misleading. The best analogue of natural selection is a cybernetic mechanism; it transmits "information" about the state of the environment to the genotype.

The point so central that it must be pressed is that natural selection is in a very real sense creative (for more about this see Chapter 6). It brings into existence real

novelties—genotypes which never existed before. Moreover, these genotypes, or at least some of them, are harmonious, internally balanced, and fit to live in some environments. Writers, poets, naturalists, have often de-
claimed about the wonderful, prodigal, breathtaking in-
ventiveness of nature. They have seldom realized that they
were praising natural selection. But a creative process runs
the risk of failure and miscreation; this is where creation
is less safe than mass production from a set pattern. Miscreation in biology is death, extinction. Paleontology
abundantly shows that extinction is the most usual end of
evolutionary lines. Some biologists who thought that natural selection is too soulless and mechanical believed instead
in orthogenesis, a theory according to which evolution is
merely an uncovering of the preformed and predetermined
organic configurations (see Chapter 6). Yet, with orthogenesis evolution would create nothing really new, and
extinction would be a mystery or plain nonsense.

Another aspect of the issue of creativity is the problem
of determinism. This will also be dealt with in Chapter 6,
and a few remarks will have to suffice here. Etkin (1964)
asks why only man was able to achieve the mode of adaptation by culture: "What did man have in his ancestry which
no other animal had that enabled him to solve his evolutionary stress?" The answer is that the human mode of
adaptation is one of the many that were possibly open to
our remote ancestors, and which ones of these modes were
to be adopted in the different evolutionary lines of the
primate order was not predetermined. It was an evolutionary invention which did not necessarily have to be made.
There are, broadly speaking, two kinds of interpretations
of evolution. One kind supposes that any and all evolutionary changes that ever occurred were predestined to
occur. The other kind recognizes that there may be many
different ways of solving the problems of adaptation to the
same environment; which one, if any, of these ways is in
fact adopted in evolution escapes predetermination. It is

because of this lack of predestination that I am inclined to question the belief that, if life exists in different parts of the universe, it is bound to result in formation of manlike, or perhaps even of supermanlike, rational beings. Together with Simpson (1964) and Blum (1965), I consider this not merely questionable but improbable to a degree which would usually mean rejection of a scientific hypothesis.

4

Self-Awareness

and Death-Awareness

There is no more succinct, and at the same time accurate, statement of the distinctive quality of human nature than that of Dostoevsky: "Man needs the unfathomable and the infinite just as much as he does the small planet which he inhabits." Gardner Murphy (1958) gives a more biological but a negative formulation: "The human nervous system possesses, then, curious and profound hungers for many objects which are neither meat nor drink, neither satisfiers of oxygen need, nor of sex need, nor of material need, nor any other more obvious visceral demand." What is "curious" about these needs is their origin in biological evolution. We know that biological evolution is utilitarian; evolutionary changes occur in response to environmental challenges, and natural selection makes the changes favor the perpetuation of the species in which they occur. The changes serve to maintain or to advance the adaptedness of the organism to its environments, i.e., the ability to survive and to leave progeny in these environments. The hungers for "the unfathomable and the infinite" do not seem, however, to promote adaptation; they are, in fact, liable to damage the prospects of survival of those in whom they are especially strongly developed. And yet it is indisputable that the potentiality to experience such hungers is biologically, genetically, implanted in human nature. The proof is that these hungers occur only in man; even the cleverest animal cannot imagine the infinite and, therefore, cannot hunger for it.

It has been pointed out above that useless and even harm-ful qualities may sometimes become established in evolu-tion, provided that they are components of organic systems which are adaptively useful as wholes. One can hardly imagine anything so preposterous biologically as the pain-fulness and hazardousness of human childbirth. This fla-grantly maladaptive quality is, however, a part of the com-plex, including also such things as erect posture, larger heads and brains, long infancy and childhood during which the individual acquires the cultural heritage of the group to which he belongs, human family organization, etc. (see page 57). The complexity calls for caution in attempts to arrive at evolutionary interpretations. It is a misconception to as-sume that every trait and quality of the organism must have a biological adaptive value by itself, taken in isolation from the rest of the living system. We must be reminded that what survives or dies, reproduces or remains childless, is a living individual, not a separate part of the body or an iso-lated gene. Of course, the performance of some one organ or a function may under certain conditions decide an indi-vidual's fate. The adaptive value, the Darwinian fitness, of a class of individuals is, however, a matter of a balance of advantages and disadvantages, strengths and weaknesses, abilities and failings.

A biologist cannot help being puzzled by the scant atten-tion devoted to and by the meager success achieved in psy-chology in the task of providing meaningful accounts of the differences in mental ability between man and animals. There is, to be sure, no shortage of words used to describe man's supposedly unique attributes. Man is said to be a being who lives by reason rather than by instinct; man is aware or conscious of his self; he has a mind, an ego and a superego; he is capable of insight, abstraction, symbol for-mation, symbolic thinking, and of using symbolic language. The dominant trend in academic psychology, which at least until recently was in the United States behaviorist psychology, rejects these words and the concepts for which they stand. These words are deemed to be too vague, imprecise, and unprofitable as operational tools in psychological research. Many recent textbooks of psychology simply do not mention these slippery words at all. The trouble is, however, that pretending that the problem of the nature of the difference between the human mind and the rudiments of mind found in animals does not exist is no help in bringing nearer a solution of this problem. When those who insist on using exclusively the most rigorous mechanistic approaches in science (sometimes nicknamed "hard" science) refuse to handle a problem, it falls by default to the exponents of "softer" varieties.

"Ethology, the comparative study of animal behavior, provides strong evidence (which it would take too much space to recount here) that something like conscious mind must have evolved a number of times in the course of the evolutionary history of the animal kingdom" (Thorpe 1965). Yet what is this conscious mind, and in what ways is human conscious mind different from that which Thorpe, one of the outstanding living ethologists, believes has arisen repeatedly in evolution? The viewpoint of Sherrington, that mind "escapes chemistry and physics," and that "natural science repudiates it as something outside its ken" has been quoted above. These statements should not be taken to mean a denial of the reality of mind, which Sherrington knew to be inescapable. *"Cogito ergo sum"* is one certainty that even Descartes found impossible to doubt. Sherrington was saying simply that, since mind is something experienced rather than observed, its knowledge can be derived mainly from experiencing rather than from observing. Another eminent physiologist, Herrick (1956), set forth the same idea even more explicitly. He equates "mind" with "awareness." Awareness cannot be reduced to chemistry "or adequately described scientifically by ever so complete an explanation of the mechanism employed." Awareness is self-awareness, and it "can be apprehended only by the particular individual concerned, by whom it is recognized introspectively." It is nevertheless not some kind of vital force; it is an organismic phenomenon:

The body makes the mind, but the mind is not a product made by the body as gastric juice is made by the stomach. It is body in action, a peculiar pattern of action of a special kind of bodily apparatus, just as walking is another pattern of action of a different kind of apparatus.

Penfield and Roberts, brain physiologists and surgeons, are more specific (1959):

Today it is just as difficult to give an adequate definition of the mind as it ever was. Consciousness is an awareness, a thinking, a knowing, a focusing of attention, a planning of action, an interpreting of present experience, a perceiving. These words are descriptive, but they hardly constitute a satisfactory definition.

In his A. S. Eddington Memorial Lecture, the eminent nerve physiologist Sir John Eccles (1965) raises an additional thorny problem. His starting point is that "We may take it as certain that my conscious self depends uniquely on my brain and not on other brains. . . . Two questions may be asked: What is the nature of this consciously experiencing self? And how does it come to be related in this unique manner with a particular brain?" The uniqueness of the self cannot be derived wholly from the uniqueness of the gene complement of the individual, since identical twins have the same gene complements but each of them experiences his conscious self as distinct from that of his cotwin. On the other hand, environmental influences and experiences of a lifetime, though they may modify the characteristics of the self, do not prevent the continuity of the memory of the individual down to its first awakening in childhood.

Sir John then quotes approvingly the old speculation of

H. S. Jennings, to the effect that there may exist a limited store of selves ready to attach themselves to a "substratum" of two united sex cells, and concludes that "we come to the religious concept of the soul and its special creation by God." Now, it seems to me that there is as much ground to see divine action in the formation of a human self as in any natural event. The existing variety of selves is an infinitesimally small fraction of the potentially possible ones, since the number of persons who ever lived or are likely ever to live is vanishingly small compared to the power of the sexual process to engender ever novel gene patterns. And it is also evident that another person, regardless of his genes, would feel himself different from me, even if it were conceivable that his lifetime experiences were exactly like mine. For it is the more or less unbroken continuity of experiences, remembered and forgotten, derived from the circumstances of my life and conditioned by my genotype, that makes me a self unmistakably different and nonrecurrent from any other person alive, having lived, or to live in future.

With some notable exceptions, psychoanalytically oriented writers have devoted little attention to the evolutionary origins of the human psyche. In the classical Freudian triad of id-ego-superego, the first member is derived from biologically determined impulses of man's animal nature, while the latter two are more or less distinctively human. "Where id was, there shall ego be," is Freud's pithy formulation. Menaker and Menaker (1965) present a biologically more coherent view. In their words, "When consciousness evolved as the highest expression of the evolution of nervous control, it was inevitable that an integrating principle evolve in conjunction with it. Man could not have experienced awareness without having evolved a way of organizing, integrating, co-ordinating the 'different awarenesses,' otherwise his psychological state would be chaotic." This organizer is the ego, which "is that psychological capacity through which consciousness is organized and integrated, through which the person is set in function both physically and mentally, and through which adaptive thought and behavior is achieved."

Existentialists affirm that it is man and man alone who truly "exists." This "existence" is what sets him apart from all other forms of life. The existentialists are perhaps on the right path toward identification of man's basic biopsychological singularity. However, because man experiences, strictly speaking, only his own individual consciousness or self-awareness, it is difficult to prove rigorously the existence or the nonexistence of some elements of awareness in animals other than man (cf. pages 27–28). However, to an evolutionist the most satisfactory statement, incomplete though it is, is that given by Fromm (1964). In his words:

Man has intelligence, like other animals, which permits him to use thought processes for the attainment of immediate, practical aims; but man has another mental quality which the animal lacks. He is aware of himself, of his past and of his future, which is death; of his smallness and powerlessness; he is aware of others as others—as friends, enemies, or as strangers. Man transcends all other life because he is, for the first time, life aware of itself. Man is in nature, subject to its dictates and accidents, yet he transcends nature because he lacks the unawareness which makes the animal a part of nature—as one with it.

Self-awareness is, then, one of the fundamental, possibly the most fundamental, characteristic of the human species. This characteristic is an evolutionary novelty; the biological species from which mankind has descended had only rudiments of self-awareness, or perhaps lacked it altogether. Self-awareness has, however, brought in its train somber companions—fear, anxiety, and death-awareness. Menaker and Menaker (1965) have stated this discerningly:

In the animal world from which we emerged, anxiety—or shall we say, fear—serves a survival function and appears as a warning of impending danger to be reacted to with the full panoply of automatic instinctual equipment which is available for the individual's survival. Human evolution poses a new problem, although it is motivated by the same survival need. It is obvious that the great human evolutionary acquisition, awareness, must add a special dimension to fear.

Man is burdened with death-awareness. A being who knows that he will die arose from ancestors who did not know.

Viewed in evolutionary perspective, self-awareness is primary and death-awareness is secondary; death-awareness is a bitter fruit of man's having risen to the level of consciousness and of functioning ego. Self-awareness has developed as an important adaptation; death-awareness is not obviously adaptive, and it may be biologically detrimental. Their appearance, however, bears out the view of the historian Erich Kahler (1964), who wrote:

When we survey the course of history, indeed of evolution, we cannot fail to notice the gradual expansion of existential scope, and certain events (often covering extended periods) which constitute caesuras, or turning points, because their center of gravity of existence shifts from one level to another level, from a lower level with narrower scope to a higher level with wider scope. Such a caesura was for instance the evolutionary transition from the animal to the human being.

The evolutionary adaptive significance of self-awareness lies in that it serves to organize and to integrate man's physical and mental capacities by means of which man controls his environment. According to Hallowell (1961), "What occurred in the psychological dimension of hominid evolution was the development of a human personality structure in which the capacity for self-awareness, based on ego functions, became of central importance." This personality structure is a complex including symbolic thinking, communication, symbolic language, intelligence, and reasoning, and indirectly tool using, toolmaking, and culture (Roe 1963).

Death-awareness is a concomitant of self-awareness, but

while the latter is open more to introspection than to outside observation, the former leads to forms of behavior the
consequences of which are more easily observable. One
such behavior is burial of the dead. Ants, bees, and other
colonial insects consistently remove from their nests dead
members of the colony, together with food refuse, empty
cocoons, and other debris. This is, however, only a part of
keeping the nest clean, an instinctive sanitation measure
and not anything like a sepulture. The alleged existence of
"cemeteries," to which old or weak individuals purportedly
repair to die, is questionable, and in any case bears no re-
semblance to the living practicing burial rites for dead
members of their own species.

The evidence is conclusive that the ancient race, the
Neanderthal subspecies of man, buried their dead. According to Hawkes and Wooley (1963), the most ancient known
burial is that in Wadi el Mughara, in Palestine. Not only
were several bodies laid in graves cut in the cave floor, but
they were accompanied by offerings of food and weapons.
The Neanderthal youngster in Teshik-Tash, in Turkestan,
was placed in a slab-lined grave, and his head was sur-
rounded by six pairs of horns of mountain goats. A number of Neanderthal graves have been discovered in Europe,
and the conditions in some of them definitely suggest "cere-
monial" burials. The bodies are laid in an attitude resem-
bling sleep, bedaubed with red ocher, and accompanied
by stone implements and remains of food. The red ocher
is believed to have been used to create an illusion of the
body having life-giving blood.

The remains of Peking Man, belonging to the species
Homo erectus, presumed to be ancestral to our own species,
Homo sapiens, seem to tell a story not of burials but of
cannibal feasts. A series of skulls found in the Choukoutien
Cave were all severed from their bodies, and their basal
parts broken in such a way as to facilitate the extraction of
the brain. This may mean only that Peking Man had already developed some refined gastronomic tastes. On the

other hand, similar operations were practiced by some tribes in historical times for magical and sacramental purposes. Eating the body, and especially the brain, of a slain warrior was believed to be a way to acquire his strength, stamina, and valor.

Late Paleolithic, Mesolithic, and Neolithic burials are increasingly more numerous. Concern for the dead evidently became widespread at the dawn of humanity. It is one of the cultural universals in mankind. The forms it takes in different cultures are widely variable, ranging all the way from fear and dread of the spirits of the deceased, to invoking them as helpers and protectors. The bodies are buried under the floors of the dwellings in which their kin live, or in special cemeteries at a distance from the dwellings, or in natural or artificially made caves or tombs. Or else the corpse is cremated, and the ashes preserved or buried in special containers or scattered or thrown into water. Care may be taken to preserve as much as possible the outer appearance of the deceased by embalming or mummification, or the body may be deliberately exposed to wild animals or to carrion vultures. There is no need to review here this tremendous variety of customs, rites, and attitudes recorded by anthropologists in different parts of the world. The cardinal fact is that all people everywhere take care of their dead in some fashion, while no animal does anything of the sort. Termites and some other forms eat the cadavers, but this is more akin to sanitation than to burial. Female apes and monkeys carry for days the dead and even decomposed corpses of their infants, but wounded or ill individuals simply seek a hiding place where they are least visible to predators, and remain there until they either recover or die. There is some anecdotal evidence of elephants and other animals trying to help weakened individuals of their group, but certainly not to bury the dead ones (Thorpe 1965).

Universality of a cultural trait is no proof that it is represented in the human genotype by a special gene, or that it is due to a special instinct or to an inborn drive. A suggestion has been made (e.g., by Wissler) that there are two distinct kinds of cultural traits: cultural particulars acquired by training, and cultural universals that are inborn. This suggestion does not find much sympathy among anthropologists, nor is such a dichotomy plausible on biological grounds. Human cultural capacities can hardly be generated by so crude a mechanism as one-to-one relationships between a gene and a trait or capacity. The relationships between biology and culture are more subtle, but nevertheless real. Genes made the origin of culture possible, and they are basic to its maintenance and evolution. But the genes do not determine what particular culture develops where, when, or how. An analogous situation is that of language and speech—genes make human language and speech possible, but they do not ordain what will be said.

Only a being who knows that he himself will die is likely to be really concerned about the death of others. John Donne's genuinely great insight, "Any man's death diminishes me, because I am involved in Mankinde," is not spoiled even by having become a cliché. Donne perceived that it is chiefly because of my death-awareness that I feel myself "involved in Mankinde." Anthropologists have recorded a great variety of customs and attitudes connected with death and its sequels in different peoples. All these customs are grounded on the fundamental fact of death-awareness, which is, indeed, one of the basic characteristics of mankind as a biological species.

This is not to suggest that death-awareness is a primary, irreducible, or unitary genetic or psychological entity. As pointed out above, it is probably an outgrowth and a necessary concomitant of self-awareness. To say that self-awareness and death-awareness are genetically conditioned does not, however, imply that they are simple genetic units. There is no such thing as a gene for self-awareness, or for consciousness, or for ego, or for mind. These basic human capacities derive from the whole of the human genetic endowment, not from some kind of special genes. They arise in the development of a human individual as parts of the cybernetic system of feedbacks which results in growth and maturation, both physical and psychological, of the human being. The cultural and behavioral traits which anthropologists and psychologists study are at several removes from the genes which occupy the attention of biologists and biochemists. The nature and origin of these traits are, however, legitimate concerns to evolutionary biologists, just as they are to anthropologists, psychologists, humanists, historians, and philosophers. These approaches should be complementary, not rival.

A biological question suggests itself and must be dealt with at this point. We have said that the self-awareness, and also the death-awareness which is one of its products, are genetically conditioned. If so, could they have arisen and developed biologically, through the action of natural selection, as aids to the so-called instinct of self-preservation? In particular, could the acquisition of the awareness of the inevitability of death have been an independent event, because it was advantageous in a biological sense? I think that this latter question must be answered in the negative. Death-awareness can be advantageous if it supplies motivation to a parent to make provisions for the maintenance and welfare of his progeny in anticipation of his demise. This presupposes, however, a level of mental capacity and of cultural development which is hardly conceivable without something at least approaching the human mentality as it now exists. Self-awareness is probably antecedent to, and a precondition of, death-awareness. There is no question, of course, that recognition of certain things as dangerous to life was enormously important from the earliest stages of human biological and cultural ascent. An adroit hunter provided sustenance for himself and for his progeny, while a maladroit or stupid one lost his life and let his offspring starve. But of what conceivable advantage could it have been to a primitive hunter to know that he would die sooner or later, regardless of his skill, prowess, or valor?

There is much confusion about what the "instinct of self-preservation" really is. It should not be imagined that the organism "knows" through some uncanny perceptivity what is good for it to do in order to remain alive. True enough, all organisms, from the lowest to the highest, react to stimuli that commonly occur in their habitual environments in ways that tend to maximize the chances of their survival. The key words in the foregoing sentence are, however, "that commonly occur" and "habitual environments." What the great physiologist Cannon described as the "wisdom of the body" supplies some of the fascinating chapters in biology, but the body is wise chiefly under conditions which the biological species to which it belongs encountered in its evolution. Placed in novel conditions, the body loses its wisdom, and often becomes surprisingly stupid. Excepting the albinos, our bodies react to sunlight by developing a protective suntan, but they react to X-rays and atomic radiation by radiation sickness. Our ancestors were often exposed to the dangers of sunburn, but not to those of high-energy radiations. In short, there is no instinct of self-preservation, if by such an "instinct" one means an ability to react to all environmental stimuli always in such a manner as to preserve the individual's life. There is instead a variety of responses, and among them most of the responses to stimuli which occurred frequently in the ordinary environments of the species are adaptive. Their adaptive outcomes were organized by the action of natural selection in the process of evolution.

In his later years, Freud came to believe that, in addition to an instinct of self-preservation, man has also a death instinct, or death drive. His dichotomy of Eros-Thanatos, life and death, preservation and destruction, may give a fine metaphorical description of certain aspects of the human psyche. As a biological theory, this simply does not stand up. It is true that in most living beings the development of an individual tends eventually toward senescence, decrepitude, and death. Only some of the simplest forms of life, which reproduce by simple fission of their bodies into two or more new individuals, are said to be potentially immortal. The establishment in evolution of a drive toward self-destruction and death is, however, highly improbable. This could conceivably occur only if the survival of individuals who have completed their reproduction and care of the progeny were actually disadvantageous to the reproductively active part of the population. If, for example, the post-reproductive individuals consumed the food and thus made the younger ones starve, or otherwise impeded or handicapped the young, then natural selection might possibly favor the tribes or subpopulations in which death followed closely the cessation of reproduction. It is most unlikely that this was the case in man, especially when mankind started to acquire culture. The old were the repositories of information and wisdom, in addition to their possible use as baby-sitters.

It would certainly be wrong to suppose that the forebodings of death haunt the thoughts of all people continuously. Individual variations and nuances are innumerable. Complete unconcern and coolly rational acceptance of the inevitable are feigned probably more often than really felt deep down. Continuous preoccupation and anxiety are, however, symptoms of abnormal morbidity. Understandably, the old are on the average more acutely concerned and conscious of mortality than the young, the sick more than the well, and those who have ample leisure for brooding thoughts more than those engaged in intense activity. The two valuable compendia by Choron (1963, 1964) ably review and summarize the diversity of attitudes and of philosophical musings and speculations about death. In children, at least in Western culture, the awareness of the inevitability of death develops rather slowly, starting at about the age of six but not becoming fully established until the age of eight or nine or even later.

Some students of primitive peoples questioned the universality of death-awareness in certain savages. In his remarkable study, *Primitive Man as Philosopher*, Radin (1957) has the following to say about these doubts:

It has been frequently contended that all primitive people assume that no death is ever a natural one and that the only kind of speculation they ever indulge in is to discover who has caused death. Both these contentions are quite wrong. The inevitability of death and the inexorability of fate are frequently mentioned in both song and proverb, at times with courageous acquiescence, at times with petulant complaint, at times with melancholy sadness. . . . The theme of the inevitability of death pervades the proverbs and poetry of practically every tribe.

Indeed, in our own culture death has frequently been pictured as a skeleton or an old man armed with a scythe, but it does not follow that the eventual arrival of this scythe was not believed to be natural and inevitable. It is the word "natural" that has changed its meaning. Death-awareness is universal and, as we have seen above, it arose in man not later than on the Neanderthal time level. Whether it was universal at that time level is, at present, an open issue.

A meaningful picture of human psychic evolution seems to emerge from the above considerations. Death-awareness became established in human evolution as a species trait. However, this trait was not, and possibly is not, adaptive in itself. It is an integral part of the complex of human faculties, the core of which is constituted by self-awareness, capacity for abstract thought, symbol formation, and the use of language. The adaptive potency of this complex is beyond doubt. It made possible control of the environment by changes in the acquired cultures more often than by changes in the inherited genes. Other consequences are less obvious but no less important. Having become aware of the inevitability of death man has tasted the Forbidden

Fruit. This awareness is one of the sources, possibly the prime source, of ultimate concern.

Man is existence which has evolved to become conscious of itself. He knows that he is a transient episode on the stage of nature. He feels some anxiety about death, although the acuteness of this anxiety varies greatly from person to person. He asks whether his life as an individual, and the lives of other people near and far from him, are, in Whitehead's words "passing whiffs of insignificance." The urgency with which people insist on asking this question is widely variable, but probably no one above the idiot level is entirely free of wonderment in the face of this eternal riddle. Tillich (1963) thought that every human being has an "immediate awareness that he is finite and that he transcends finitude exactly in this awareness."

I do not wish to be understood as equating the fears and forebodings of death with ultimate concern, nor did Tillich make any such claim. The situation is a complex one. No consensus has emerged from studies on the origins of primitive religion. Comparative anthropology has led to the formulation of several theories, each of which may contain an element of truth. Tylor, one of the nineteenth-century pioneers of "evolutionary" cultural anthropology, stressed that primitive man saw manifestations of personified spiritual agents in living as well as inanimate nature and in human phenomena. Religions arose, in his view, out of this all-pervasive animism. McDougal assumed a special religious instinct, which Wilhelm Schmidt claimed led man to a primitive *Urmonotheismus,* from which animistic and polytheistic religions arose secondarily by a process of degeneration. Durkheim emphasized the ties between primitive religion, totemism, and social structure.

The views of Malinowski (1931) seem to me the most discerning of all. In his words: "Religion, however, can be shown to be intrinsically although indirectly connected with man's fundamental, that is, biological, needs. Like magic it comes from the curse of forethought and imagination, which fall on man once he rises above brute animal nature." One of the foci in which this "curse of forethought and imagination" converges is man's transcience:

The existence of strong personal attachments and the fact of death, which of all human events is the most upsetting and disorganizing to man's calculations, are perhaps the main sources of religious belief. The affirmation that death is not real, that man has a soul and that this is immortal, arises out of a deep need to deny personal destruction, a need which is not a psychological instinct but is determined by culture, by co-operation and by the growth of human sentiments.

Feibleman, a psychiatrist (1963), quite appropriately stresses the psychobiological aspect:

Theologies are qualitative response systems which promise survival. Irrespective of their truth or falsity (and since they conflict, no more than one of them can be true), the overwhelming statistics as to their prevalence indicates that they are necessary for some need-reduction in the human individual. The need is, of course, the need for survival, for ultimate security, for the escape from the pain of death. The human individual knows that he must die, but has thoughts larger than his fate. . . . Religion is an effort to be included in some domain larger and more permanent than mere existence.

The great world literatures abound in descriptions of successes, and even more of failures, of these human efforts. The concerns of literary art overlap those of evolutionary science in this field. Is a dialogue between the artists and the scientists possible? I know of no more poignant portrayal of an abject, but all too common, failure to achieve inclusion in a domain "more permanent than mere existence" than that given by Tolstoy in "The Death of Ivan Ilyich." An ordinary man, moderately prosperous, moderately successful, and not too unhappy in his family life, begins to realize that he is gravely ill, and that death is near.

He tries to be sensible about it; he recalls an example of syllogism from a textbook of logic: "Gaius is a man, men are mortal, hence Gaius must die." Suddenly he recoils from all logics—"that Gaius was a man, a man in general, and it was quite just; but I am not a Gaius, I am not a man in general, I have always been a being quite, quite different from all others. . . . And it is impossible that I too must die. This would be too dreadful." We do not know what a dying animal feels, but it is most unlikely to experience the terror of Ivan Ilyich; is this the evolutionary acquisition corresponding to the Biblical symbol of Fall?

A quite different sort of rebellion against life and death, in a sense the very antipode to that of Ivan Ilyich, is dissected and examined by Dostoevsky in *The Possessed*. Kirillov, the rebel, finds no sense or meaning anywhere in the whole world. Everything, including even all nature with all its laws, is a tissue of lies and a "devil's vaudeville." In the midst of this absurd world there is, however, a conscious being, Kirillov, who understands that the world is senseless. And he knows that he happens to be placed, without his consent, in the role of an actor in this "devil's vaudeville." His dialectics show him only one way to escape from his predicament with dignity. This is self-destruction. A freely arrived at decision to commit suicide is the sole and only way for a man to capture the chief attribute of divinity—limitless freedom. He kills himself, in order to assert his "new awesome freedom," and thus to become god-like. There is in all this one thing that is undeniable: Kirillov's dialectics transcend animality. Neither is he a cringing coward, like Ivan Ilyich.

For two and a half millennia, since the dawn of philosophical speculation in ancient Greece, the meaning and significance of life and death have been among the principal problems of philosophy. Only the influential modern school of logical analysis, which prides itself on being rigorously "scientific," has declared these problems to be meaningless. There is, nevertheless, something irresistibly and overwhelmingly urgent and attractive about these particular "meaningless" problems. And although science cannot claim to solve them, it can perhaps furnish some information relevant to the speculations of the philosophers.

Socrates said that "the true philosophers are ever studying death," and that a philosopher "is ever pursuing death and dying." These sayings have echoed through the ages. According to Seneca, philosophy teaches that man must live "as if he were on loan to himself, and is ready to return the whole sum on demand" (this and the following quotations according to Choron 1963). Montaigne thought that "To philosophize is to learn how to die." Spinoza hoped that a philosopher is "less subject to those emotions which are evil, and stands in less fear of death." Schopenhauer makes a statement which no biologist can disown, namely, that, "In the case of man the terrifying certainty of death necessarily entered with reason," and he continues: "All religious and philosophical systems are principally directed to this end, and are thus primarily the antidote to the certainty of death, which the reflective reason produces out of its own means." It is, however, with the existentialists that man's transience becomes really the center of gravity of all philosophical thinking. Jaspers is quite emphatic about this: "If to philosophize is to learn how to die, then this learning how to die is actually the condition for the good life. To learn to live and to learn how to die are one and the same thing."

It must be made quite clear that those who believe in some form of immortality of the soul, or in resurrection of the dead at some time in the future, are no less preoccupied with the problem of man's transience than are those who hold death to be complete and final annihilation. This is loudly proclaimed by the glorious monuments of ancient Egypt; by the paintings and sculptures in almost any Christian church (except the bare meeting houses of some Protestant sects); by Dante's great classic; by the fundamental tenets of the other great religions—Buddhism,

Hinduism, and Islam. There is, of course, no denying a fundamental difference. To a believer in resurrection and immortality, the temporal existence is merely a preparation for the existence to come; to those who hold death to be the ultimate dissolution, the meaning of life must be found in what happens between birth and death, or else no meaning can be found at all. A believer in immortality can therefore regard his personal salvation as his ultimate concern, and the purpose of his life. If there is no immortality, the problem becomes vastly more complex and difficult. A meaning of life can only be found in something greater than, but including, the personality.

5

Search

For Meaning

Aristotle called man *politikon zoon,* a social animal. Of course, man is by no means the only social animal. There are many others, and some of them are, so to speak, more emphatically social than man. The division of labor among members of the societies of termites, ants, social bees, and wasps has been carried much further than in human societies. In these insects most individuals are "workers" that have lost the basic biological function of reproduction; the workers are sterile, or, if you wish, sexless, although anatomically they can be shown to be either underdeveloped females or (among termites) both underdeveloped females and males. Reproduction is delegated to a specialized caste of individuals, fertile females and males, the former rather misleadingly called "queens," since one thing which they do not do is to serve as leaders or rulers.

Man's sociality is of a very special kind, as Aristotle clearly realized. *Politikon* implied living in city-states; man is a "political" animal. The foundation of man's sociality is self-awareness and symbolic communication; that of insect societies is inherited instinct. The basic social unit in man is a nuclear family, an association of mother, father, and children. Although the customary relationships among members of a nuclear family are subject to endless variations in different cultures, the nuclear family is very nearly a cultural universal in mankind.

Where and how the nuclear family arose in the evolutionary history of the human species is conjectural. In its human form it is not found among the nonhuman primates, which exhibit, however, an impressive variety of social organizations. Primate behavior, reproductive physiology, and child-rearing practices have in recent years been the subject of extensive and ingenious studies (see the symposia edited by Roe and Simpson 1958; Washburn 1961, 1963; Southwick 1963; Etkin 1964; De Vore 1965; and the books of Oakley 1961, Schaller 1963, and others). The size of a social group varies from 2 to 5 (in orangutans) to from 13 to 185 (some baboons). In general, the groups tend to be larger in species that live and feed on the ground than in arboreal ones. A troop usually consists of several to many sexually mature females, one or several mature males, and young who have not yet reached the reproductive stage. Among the mature females, some are sexually receptive, others pregnant, and still others nursing infants. Excepting the infants being nursed, each animal forages for food for itself. There is little or no cooperation in most species in the matter of procuring nutrition. There is, however, cooperation for mutual defense; in particular, the strongest males take defense positions when the troop is in danger, especially from predators such as large cats.

There has been some difference of opinion as to the amount of fighting that occurs among members of a troop. When baboons are confined in zoos, there may be fierce combats among males procuring and defending their "harems." Observations in nature show, however, something quite different. There is a more or less strict order of dominance and subordination among the individuals of a troop, but this order is established and maintained usually by ritualized threats rather than by actual fights. In some species the weaker sexually mature males are excluded and form "bachelor" bands hovering in the vicinity of the troop, awaiting their chance. The dominant males have most access to sexually receptive females, but it is by no means true that the whole offspring in the troop is fathered by a single male. On the contrary, there are usually several active males, and in some species a considerable sexual "freedom" seems to prevail. The most important difference in sexuality between man and his relatives is, however, something else. Among the primates other than man, a female is sexually receptive only during rather limited periods of oestrus (heat). The human female, when mature, is continuously receptive.

In a monkey band, though only a minority of the females are in oestrus at any given time, there may be one or several females available in this state. The sexually active males thus have more or less continuous access to receptive females in the troop, but these are different females at different times. This makes unlikely a monogamous attachment and paternal care of the children. In fact, the males in a monkey band do, as stated above, defend the immature individuals against attacks of predators, as they defend also the other members of the troop. In some species at least, the males are, however, at best only tolerant of the young, and at times these latter have to take care to keep out of the way of their fathers and uncles. In man, the continuous sexual receptivity of the female serves to reduce the competition among the males for mates and makes monogamy practicable.

The prolonged helplessness of human infants imposes a far greater burden on their mothers than is the case with the rather more nimble babies of monkeys and apes. A human female when she becomes a mother is likely to be no longer able to procure adequate food for herself and her offspring. This is especially true when the food supply is obtained by hunting game, rather than by collecting vegetable products, insects, or other small animals in the near vicinity of the camp. The human male perforce must take upon himself the role of a family provider. As Hockett and Ascher put it in the felicitous phrase quoted above (page 57), "All this makes for the domestication of fathers."

The family is the oldest and in all probability the most permanent of human social institutions. It is "permanent" not in the sense that it does not change, for there are countless variations in family structure, but rather because as a basic institution it will exist in some form probably as long as the human species endures. There has been much debate concerning the relative importance of genetic and cultural determinants of human sexual and familial dispositions. Both are manifestly involved. That the sexual drive is the outcome of genetically established physiological mechanisms will probably not be doubted by anyone. The manifestations of this drive are, however, conditioned by the cultural setting. It should not be supposed that because it is species-wide, the genetic basis of human sexuality is uniform everywhere. On the contrary, genetic variations in the intensity, the timing, and in the preferred kinds of sexual gratification are quite frequent and important. The complexities of customs and of other sociocultural factors determine the overt behavior and the forms into which the sexual desires are channeled. A tremendous superstructure of precepts and taboos has become imposed upon the genetically conditioned physiological drives. Some societies are more permissive and others more restrictive in the rules which they establish to regulate the sexual, family, and progeny relationships. But no society, past or present, of which we have a record, left these matters entirely to individual caprice or judgment.

Genetic and cultural factors are intertwined also in the parent–child relationships. Although this matter has been quite insufficiently studied from the point of view of genetics, it is highly probable that the focusing of the emotional attachments of the mother on her child or children is rooted in the genes. The same is probably true of the father–child relationship, although here the genetic roots are more shallow, and anthropologists have described a greater variety of cultural situations.

The importance for human development of the helplessness of the human child and of its complete dependence on its mother can hardly be exaggerated. For the species to survive, evolutionary adjustments had to occur, and these adjustments are of basic biological as well as cultural significance. As pointed out by Waddington (1960), a child has a genetically established capacity to become an "authority acceptor" and an "ethicizing being." The evolutionary process has not provided man with set ethical principles and values, but it has equipped his children with an inclination to absorb such principles from his parents, relatives, and other carriers of authority. This facilitates the transmission from generation to generation of culturally evolved ethics and values.

Many biologists, including some would-be "humanists," have liked to speculate that human ethics have become shaped and fixed by natural selection as genetic instinctoid drives. The real situation is much more interesting. What is established as a biological adaptation is the ability to "ethicize," not the nature or the contents of the ethical tenets. These latter come from the cultural, not from the biological, evolution. In Freudian terms it can, I suppose, be said that what is biologically given is a capacity to develop a superego, not the kind of superego that develops. Those who might argue that the biologically given basis of the superego is therefore unimportant should be reminded that there seem to exist both eager and refractory authority acceptors, and that these variations may well be in part genetic.

For man, parenthood acquired, presumably in the evolutionary early stages of humanity, a significance which for an animal it cannot have. This comes from the human self-awareness and death-awareness. The parent is aware of his transitoriness. Children, grandchildren, and great-grandchildren are likely to live when the parent is no longer living. Man has a hope, perhaps an illusory one, that he somehow survives in his descendants. A life devoted to one's family and to one's progeny (biological or even adopted) seems to acquire a meaning; it may be experienced as capturing a particle of an immortality which is beyond the reach of an individual.

Fertility symbols are among the most ancient evidence of man's spiritual awakening. According to Hawkes and Wooley (1963),

What in fact provided the strongest and most definite bond between the Palaeolithic and Neolithic religious impulses sprang from their common desire for fertility. It has been seen that although Palaeolithic religion may have been very much bound up with totemistic animal cults and hunting magic, the most developed and clearly defined cult objects to have survived are the Mother Goddess figures and carved phalli. In the primary Neolithic cultures these two fertility symbols are still absolutely dominant. . . . However much is uncertain, nothing can shake the mute evidence of hundreds upon hundreds of little clay, bone and stone effigies of the Mother Goddess.

To our modern tastes these ancient "Venuses," with their pregnant bellies and overdeveloped buttocks, may seem ugly or ridiculous. Fertility cults, more or less elaborately disguised, are nevertheless alive today. Phallic symbols are still prominent in some religions, both primitive and advanced ones. A conventionalized phallus and vulva are symbols of the god Siva in Hinduism. A curious attempt to interpret the sexual act and human procreation as quasi-religious sacraments was made by the Russian philosopher Rozanov (see in Zenkovsky 1953). Related ideas have been broached by Goodenough (1965).

In a larger sense, it may be said that modern variants of cults of fertility supply the meaning of existence to many millions of persons now living. Man (or woman) strives for sexual gratification, then for family attachments, and finally for the security and welfare of the progeny. These strivings form designs for living which are so firmly anchored in the genetically established instinctoid drives that their meaningfulness is taken for granted by almost everyone and is questioned by few. Although undeniably biological in their roots, these strivings on the human level easily take on the cultural elaborations and embellishments of symbolism, myth, mysticism, poetry, and art. Motherhood is esteemed as meritorious or even sacred by most peoples. The Madonna and Child became hallowed symbols of Christianity. We need not go so far as to see here the ancient Mother Goddess who has reclaimed her ancient dignity; the deep emotional appeal of the mother image is, however, evident.

Man is assuredly not unique as an animal that seeks not only to mate but also to cooperate with his mate in bringing up the progeny. This cooperation makes excellent sense biologically, and it has evolved independently in different groups of animals. For example, in many species of birds the female and male parents are not only monogamous for a breeding season but also work together in feeding and protecting their offspring. Might this be one of the reasons why bird-watching has such a fascination for so many people? Even self-sacrifice by the parents on behalf of the progeny is not a great rarity; in some species of birds and mammals, a parent places himself in a position of danger from an enemy's attack, to shield his young from peril. Whether this objectively self-sacrificial behavior is identical with human altruism is a different problem. Many observers have felt that it is, but it may be doubted that altruism can really exist without an ability to envisage alternative courses of action and freely make one's choice among them. On the other hand, parental care is quite certainly not limited to man. However, it is only in man that the comparatively weak instinctoid drive is reinforced by human strivings for relatedness and love. Therefore, it is also in man alone that parental devotion is liable to yield a disappointment, a parent meeting alienation and rejection instead of love in his children.

May one, then, be satisfied to live to be alive, to be alive to leave more life, life which will continue the chain of living, one hopes forever? To many people, this is the only plausible answer to the problem of the meaning of life, and the endless chain of living the only believable kind of immortality. If it were possible for an animal to ask what is the meaning of its life, the only answer would be that the meaning of life is life itself. But is this answer good enough for man? An animal capable of asking such questions would have to be human!

Camus saw in a flash of insight: "I know that something in this world has a meaning and this is man; because he is the only being that demands to have a meaning." Herein is the strength and the weakness of the ageless idea that life is an end in itself. This idea sustains life; can it also satisfy a mind blessed, or cursed, with self-awareness and death-awareness? Must a meaning of human life be found in something beyond and above it, if it can be found at all?

Man had always to toil and to struggle to keep himself alive, before asking the question what he was living for. Some of us, living amidst material plenty and relative safety, fail to realize that the business of keeping ourselves alive was for most people, and for many still is, a full-time occupation. Hunger was never far off, and diseases and calamities of all sorts were perpetually at man's heels. Those who have experienced living under such conditions know that the drive to survive may be intense and unwavering, and yet this drive does not entirely displace the question of the sense and meaning of it all.

Following the lead of Tylor and Frazer, many anthropologists have held that in the struggle to remain alive man invented magic, and from magics there developed religions. This view is not shared by all students of these problems, and many assume independent origins for magic and religion. Religion is what man does with his ultimate concern; magic is a utilitarian activity. Baffled by the forces of nature that he does not understand, man wishes to coax or even to command them to do his bidding. This utilitarian function of magic is described by Malinowski as quoted on page 78. Magical beliefs and practices cling to probably all religions, primitive as well as advanced ones. We need not enter here the disputed ground and decide whether magics can sometimes be effective. One need not be a supernaturalist to recognize that magics occasionally do accomplish something like psychological miracles for those who believe in them.

Goode (1955) rightly insists that

we are required, then, to grant equal intellectual and scientific importance to the religious systems of the Dahomey and the Augustus Romans, the Zuñi and the Tibetans, the Lacandones and the Unitarians, in short, to the primitive and the "civilized" religions. Now, it is true that certain religions have had greater impact upon our personal lives, because they form part of our group tradition—the Hebrew, for example. This fact does not make them more fundamental than primitive religions in our research into the major structural facts about religious systems.

The same is true of the aspect of the problem which occupies the attention of an evolutionary biologist. The human species has developed a capacity, in fact a need, to have a religion; what could have been the stimulus, or the function, which made this capacity one that increased the fitness of its possessors? This is a meaningful problem even if Goode is right that "The religions of primeval man will remain forever unknown to all social scientists, though many poet-scientists will persuade themselves that they have intuited these meanings."

Magics, no matter how earnestly they are believed in, provide no insight into the meaning of existence. Whether or not they are ever useful for survival, they fail to tell us what we survive for. Religions do provide answers, the validity of which is, of course, a separate problem. Man is in the world through God's will, and his meaning is to be God's servant. God or gods, no matter how differently they are envisioned in different religions, are immortal and eternal. Man is aware of his transitoriness and fragmentariness. Hartshorne describes religion as "man's acceptance of his fragmentariness." Man overcomes his transitoriness and fragmentariness by becoming, at least in his imagination, a part of the sublime and eternal life.

The form of participation in the eternal life is conceived differently by different religions. In the Old Testament there was no promise or threat of paradise or hell, but a doctrine of the resurrection of the dead became established in Judaism in post-Biblical times. This doctrine is explicit in Christianity and Islam. The fundamentalist conception has been represented, often with dreadfully realistic details, in Christian churches, in thousands of paintings showing the Last Judgment; it is described with the same attention to frightful details in Dante's great classic. Death is, then, merely a way station in the endless career of a human soul. Death will eventually be superseded by resurrection, and not of the soul alone (which never really dies) but also of the body, which will then exist forever. The life everlasting is, however, appallingly dichotomous; it has only two alternatives—either the infinite bliss of paradise as a reward for good deeds, or endless sorrow and torment in hell, the retribution for misdeeds.

Buddhism and Hinduism reject this kind of immortality altogether. The soul freed from the shackles of the body at death is reincarnated, here on earth, time and time again, in other bodies. The succeeding reincarnations may be nobler or ignobler than the preceding ones, depending on the merits or demerits accumulated during the previous lives. These wanderings of the soul through repeated incarnation cycles are regarded as not good or desirable. Quite the contrary: the final goal of spiritual advancement is to bring these peregrinations to an end, and to achieve ultimately a reabsorption of the individuality in God or in the primal essence of the Universe-Brahma or Nirvana.

Creative thought in Christianity is seemingly moving toward a kind of synthesis of the opposing views concern-

ing immortality. Modern man is revolted by the injustice
of the dichotomy paradise–hell, the saved and the damned.
The integrity or depravity of human lives falls obviously
in a continuum, not in two discrete,classes. Tillich (1963)
attempts to resolve the difficulty as follows:

Where the symbol of immortality is used to express this popular superstition, it must be radically rejected by Christianity;
for participation in eternity is not "life hereafter." Neither is
it a natural quality of the human soul. It is rather a creative
act of God, who lets the temporal separate itself from and
return to the eternal.

And further:

The symbol of resurrection is often used in a more general
sense to express the certainty of Eternal Life rising out of the
death of temporal life. In this sense it is a symbolic way of
expressing the central theological concept of the New Being.
As the New Being is not another being, but the transformation
of the old being, so resurrection is not the creation of another
reality over against the old reality but is the transformation of
the old reality, arising out of its death.

In every known human society, in every culture of the
past that has left a historical record, peoples have arrived
at some system of religious views concerning the meaning
and the proper conduct of their lives. Although different
religious systems are not alike, and at some points are incompatible, they perform indispensable functions. Religion enables human beings to make peace with themselves
and with the formidable and mysterious universe into
which they are flung by some power greater than themselves. Since remotest antiquity, religion has been a cultural
universal in mankind, because its symbols, myths, and phi-
losophies provided answers to the ineffable problems of
human existence. And it is because religion provides, or
seems to provide, these answers that it has served as a social
cement.

It is often claimed that religion was socially and even biologically useful because it gave supernatural sanctions to ethics; this useful function has ended, or will end, when people no longer regard such supernatural sanctions as necessary. Religion was able to supply such sanctions only because it gave meaning to human actions in general, by providing for these actions a referent in man's ultimate concern. The lives of individuals, and, even more, of the societies of which they are members, are seen as parts of a grand design enjoined by the divine power. From time to time this idea has been exaggerated to provide sham justifications for religious wars and persecutions. However, it has also led people to devote their lives to working for their fellow men in the spirit of charity, humility, and love, with the assurance that by so doing they are working for God. Religion thus facilitates the acquisition by individuals of an identification with groups larger and presumably more enduring than one's offspring or one's nuclear family.

A tribe, a nation, and finally mankind as a whole will abide beyond the span of an individual's life, even unto eternity. So, it is to be hoped, may an enterprise, such as science or art, a movement, such as a church or a sect, or an institution, such as a political party. Time after time, people have shown themselves willing to sacrifice their lives for the real or imaginary benefit of charismatic "causes" which continue to exist and to grow when the individual who sacrifices himself is dead. The human ability to acquire identification with groups, movements, and institutions confers cohesion upon them. The importance of this cohesion in human evolution is very great. It makes human history different from biological history. The crucially important point is that the group cohesion in man is predicated upon his self-awareness. The cohesion of a herd, a flock, a beehive, or an anthill stems from instinctual behavior handed down through genes.

Regardless of one's religious or agnostic preferences, it must be admitted that religion is an essential part of culture. Toynbee (1961) goes even further, since he subscribes to the opinion of another historian, Dawson, that "The great civilizations of the world do not produce the great religions as a kind of cultural by-product; in a very real sense, the great religions are the foundations on which the great civilizations rest." Be that as it may, a part of a culture cannot be discarded without doing violence to the whole. As stated by Mead (1955),

Whether a culture be tightly integrated or flexibly and loosely integrated, whether its ethics are internally consistent or profoundly contradictory, whether practices in one area of life are replicas or reversals, or, when superficially inspected, apparently unrelated to other areas of life, they will be found, upon close anthropological inspection, to be related in systematic ways.

Western civilization (including that of Russia, termed by Toynbee the Orthodox Christian civilization) has evolved on the Judeo-Christian religious foundation. For more than a millennium Christianity has supplied the meanings and aims of individual existences and of group enterprises. It is at the very core of our cultural tradition. Yet starting about a quarter of a millennium ago, and with increasing impact thereafter, this religious foundation has been eroded away, owing to its real or apparent conflicts with the growing power of scientific discovery. Literally thousands of books and articles have been written about these conflicts, some attempting to reach a so-called "reconciliation" of religion and science, others trying to show that it is science alone that can provide mankind with a spiritual foundation for its existence, thus making all other religion superfluous. It is not my intention here to review, or to summarize, even briefly, this confused and confusing literature. I do think that science, and particularly biology, are relevant to man's ultimate concern. To elucidate this relevance, one must be reminded of certain facts and of some attempts to fit these facts into the scheme of things.

Copernicus was on his deathbed when his book, showing that the earth is not at the center of the universe but is merely a planet rotating around the sun, was published in 1543. The book attracted at first relatively little attention. Calvin rejected its conclusions with the contemptuous remark: "Who will venture to place the authority of Copernicus above that of the Holy Spirit?" But matters were not allowed to rest there. Kepler in 1609 and 1619, and Galileo in 1610 and 1632, showed that Copernicus was indeed right. This raised a storm; Galileo was summoned before the Holy Inquisition and forced to abjure his heresy. In 1637 came Descartes's *Discourse on Method,* in 1682 Newton's *Principia Mathematica,* and in 1690 Locke's *Essay Concerning Human Understanding.* In his old age, Newton saw fit to describe himself as a collector of bright pebbles on the shores of the great unknown. Nevertheless, his picture of the universe was so generally accepted that religious authorities became wary of opposing his findings, as they did oppose those of Galileo. The unseemly performance was nevertheless repeated about two centuries later, in 1859 and thereafter. Darwin's theory of biological evolution was brushed aside, because it allegedly contradicted what some people held to be the divinely revealed Biblical accounts of Creation. In 1925 the world was treated to the ridiculous spectacle of the Tennessee "Monkey Trial." At about the same time, but without noise of publicity, Teilhard de Chardin, one of the most profound religious philosophers of our time, was forbidden by his superiors in the Society of Jesus to publish his works outlining his conception of man as a product of evolution.

There are still many people who are happy and comfortable adhering to fundamentalist creeds. This should cause no surprise, since a large majority of these believers are as unfamiliar with scientific findings as were people who lived centuries ago. The really extraordinary phenomenon is the continued existence of a small minority of scientifically educated fundamentalists who know that their beliefs are in utter, flagrant, glaring contradiction with firmly established scientific findings. There is (or until recently was) published in the United States a journal of Deluge Geology, and at least one ostensibly "scientific" book dealing with this strange subject has been printed. I had a futile and exasperating correspondence with one antievolutionist "creationist," who could not be accused of unfamiliarity with the relevant evidence. Discussions and debates with such persons are a waste of time; I suspect that they are unhappy people, envious of those who are helped to hold similar views by plain ignorance.

Science and religion deal with different aspects of existence. If one dares to overschematize for the sake of clarity, one may say that these are the aspect of fact and the aspect of meaning. One can study facts without bothering to inquire about their meaning. But there is one stupendous fact with which people were confronted at all stages of their factual enlightenment, the meaning of which they have ceaselessly tried to discover. This fact is Man. Religious Scriptures (Judeo-Christian, Islamic, Buddhist, or others) are not manuals of science. They speak instead in religious symbols. These symbols

open up a level of reality, which otherwise is not opened at all, which is hidden. We can call this the depth dimension of reality itself. . . . Religious symbols open up the experience of the dimension of this depth in the human soul. . . . The discussion of ultimate reality is the dimension of the Holy. And so we can also say, religious symbols are symbols of the Holy (Tillich 1964).

This does not quite mean that man must keep his scientific and religious thoughts in separate watertight compartments. Therefore, we always face the danger clearly seen by Santayana (quoted after Murray 1960):

A religion which, like Christianity, seizes the essence of life, ought to be an eternal religion. But it may forfeit that privi-

lege by entangling itself with a particular account of matters
of fact, matters irrelevant to its ideal significance.

Such entanglements constantly do arise. The scientific
understanding of nature and of man as a part of nature is
obviously incomplete. Some natural phenomena are and
others are not yet understood. To some people, even to
some scientists, it seems tempting to fit God into the gaps
between natural events, the gaps for which no adequate
scientific explanations are yet available. The futility of this
pious stratagem has been discussed above, in Chapter 2.
Gaps can be found easily enough. Any scientist worth his
salt knows plenty of unsolved problems and phenomena
not yet understood. Are these phenomena preferential
points of God's interventions? And are the phenomena
which are scientifically more or less well understood therefore outside the province of divine activity? Nobody will
seriously undertake to defend so inane a notion. And yet
nothing gives more pleasure to a rather common type of
religious person than to point out that science cannot explain this or cannot account for that!

What is unexplained is taken to be unexplainable, although even a superficial acquaintance with the history of
science should be enough to reject this hoary fallacy. Nevertheless, from Cardinal Bellarmine, who thought that Copernicus's and Galileo's "pretended discovery vitiates the
whole Christian plan of salvation," to William Jennings
Bryan, who said that "It is better to trust to the Rock of
Ages than to know the ages of rocks," and that "The contest
between evolution and Christianity is a duel to the death,"
attempts have been made to interdict to science the access
to great areas of knowledge. The underlying reason has
always been the same—the desire to protect the religious
heritage from attacks of the skeptics and unbelievers. This
leads to a fear that if it is admitted that any part of the
religious tradition is discardable, there is no line that can
be drawn anywhere at which to make a stand. In this, the

religious leaders fail to perform their main duty. They must be able to distinguish between what is eternally true in their beliefs and the accretions that pay regard only to a certain level of popular understanding, or to a certain sociopolitical situation in which any religion finds itself during its history. The result of the stubborn standing pat has been the continuous and ignominious retreat of fundamentalist religion.

Christianity was once a religion that really pervaded the life of the people in Western civilization. In the Age of Enlightenment, the eighteenth century, the "enlightened" discovered that the hypothesis of God was a superfluous one. The universe was a vast but very precise machine, running according to the immutable laws discovered by Newton. Some of the "enlightened" still conceded that there may have been a God who devised and built the machine; however, after he set the machine running like a supremely exact clockwork, this God no longer interfered with its operation. He is something like an absentee landlord who participates little or not at all in the affairs of his estate. Clearly, there is no use for men to address prayers or supplications to this deistic clockmaker god. Agnosticism and indifferentism are at present no longer prerogatives of an especially enlightened elite. In the West, and also in the advanced countries of the East such as Japan, probably a majority of the populations pay only lip service to their traditional religions. This is true even of some of those who go to church every Sunday and who swell the statistics of religious denominations.

And yet modern man, this enlightened skeptic and agnostic, cannot refrain from at least secretly wondering about the old questions: Does my life have some meaning and purpose over and above keeping myself alive and continuing the chain of living? Does the universe in which I live have some meaning, or is everything just a "devil's vaudeville"? Strangely enough, these questions have of late acquired a greater urgency than they had because the word has got around that philosophers have decided that sensible people should not ask such meaningless questions. In consequence, ersatz religions arise and win large followings, because they claim to answer the old questions, not only without conflicting with science, but even "scientifically." By far the most important ersatz religion has Marx and Lenin as its foremost prophets. It claims that it is not based on mere faith, but that its truth is scientifically demonstrable and demonstrated. It promises to every human his equitable, though not necessarily equal, share of the fruits of the earth, and undertakes to prevent the greedy from usurping what should belong to the commonweal. It pledges to halt the exploitation of man by man and to vouchsafe to everybody a full development and realization of his individual talents, tastes, and socially useful capacities. It would have everyone live and work for the general good—for the good of those now living and for that of posterity.

Now, all this is what Christianity also holds as desirable, but what it has admittedly and conspicuously failed to achieve during the one and one-half millennia of its supremacy in Western civilization. Communism takes the social program of Christianity, promises to have it realized universally and speedily, and declares all else in Christianity a snare, a delusion, and an opiate for the exploited masses. It is for this reason that communism is justly considered a Christian heresy, "in the sense that it had singled out several of the elements in Christianity and had concentrated on them to the exclusion of the rest" (Toynbee 1961). The truncation and simplification of the religious ideas in communism, compared to their complexity and paradoxality in Christianity, facilitates its appeal to millions of people. It has become a world force, and its power is apparently still growing in the developing countries. It unmistakably inspires total dedication in some individuals, particularly in youth, and it arouses a social dynamism of unprecedented intensity.

Yet the quasi-religion of communism collapses on closer contact and acquaintance more utterly than any other religion. Nobody has indicated the reason more simply and conclusively than Pasternak: "All this [communist ideal] is still far from realization, and yet mere talk about this has already cost such rivers of blood that, really, the end does not justify the means." To the uncounted millions of murdered, tortured, and spiritually deformed Stalin's victims this talk has brought fear, forced submission, hidden dissent, anguish, but remarkably little simple human happiness.

There is nothing but *obiter dicta* in the whole communist philosophy to persuade an honest doubter that a life dedicated to serving one's fellow men is more desirable than an exclusively self-serving career. It is not a self-evident axiom that the future of mankind is so precious to us that each and every one now living must willingly devote and perhaps sacrifice himself for those remote and perhaps even hypothetical inhabitants of the future world. Most certainly we shall not be here to behold, let alone to participate in, the future bliss. Dostoevsky's Ivan Karamazov makes this unmistakable in these words:

I have suffered not in order that my torments and my misdeeds would serve to manure the soil for somebody else's harmony. No, I desire to see with my own eyes that the lion and the antelope will lie peacefully together, and that the murdered one will arise and will embrace his murderer.

The high-minded view that the meaning of life is to strive to make the world a better place to live in is, certainly, a part of many secular ersatz religions other than communism. There is no obvious reason why this noble idea could not imbue people in the democratic societies with as much fervor as it does in at least some ardent communists. The fact of the matter is that it does so only for small minorities. This is not because of any lack of propaganda or indoctrination. Politicians of all stripes mouth this idea constantly, especially in the introductions, preambles, and oratorical flourishes with which they embellish their pronouncements. It is often presented with at least a passing reference to Christianity or to other great religions as a supporting argument. The sincerity of all this is, to say the least, not free from suspicion. Living for good causes may also be commended as a secular ethic. In his book *The Case for Modern Man,* Frankel (1959) has given perhaps the most cogent and eloquent defense of this ethic as a base for what amounts to a secular quasi-religion. Neither he nor anybody else has, however, succeeded in answering Ivan Karamazov's doubts and demurrals.

That one should live decently, help others who need help, and do what one can for those who will live after us are ethics which are widely accepted, at least in theory. In the West, their historic roots are in the Judeo-Christian religious tradition, but they have now become emancipated from dependence on this religion. That these ethics can by themselves satisfy human longings to give life a meaning is more doubtful. Devastating critiques of such attempts have been given by modern existentialists. As befits good intellectuals, no two of them really agree with each other. I take Sartre as an example, fully realizing that I cannot possibly do justice to the many ideas and subtle shades of meaning of this brilliant and extremely prolific writer.

The most concise statement of Sartre's philosophy seems to me contained, not in his formally philosophical writing, but rather in the play *Les mouches.* This is a sort of a parody on Aeschylus's *Oresteia.* Sartre's Orestes defies the gods instead of submitting to their judgment. He answers Jupiter's admonition, "Orestes, I created you and I created all things," with a rebellious cry:

Alien to myself, I know it. Outside nature, against nature, without excuse, without recourse save myself. But I shall not return under your law; I am condemned to have no other law but my own. Nor shall I return to nature, where a thousand paths are marked out, all leading up to you. I can only follow my own path. For I am a man, Jupiter, and each man must find his own way.

The rebellion ends, however, in a feeling of disgust, in "nausea." There is no God, and no meaning in the universe. In a meaningless universe man has no meaning either:

And without formulating anything clearly, I understood that I had found the clue to existence, the clue to my nauseas, to my own life. In fact, all I could grasp beyond that comes down to this fundamental absurdity. Absurdity: another word.

The universe has no real beauty: ". . . The real is never beautiful. Beauty is a value which applies only to the imaginary and which entails the negation of the world in its essential structure." Evil is, on the contrary, very real. Not only real but also unredeemable; some of Sartre's thoughts about evil are almost paraphrases of those of Dostoevsky's Ivan Karamazov. Absurd and meaningless man has nevertheless a capacity for freedom. His freedom is, however, of a peculiar sort—it is the freedom to say, "No." In some, to me incomprehensible, ways, "Freedom is the human being putting his past out of play by secreting his own nothingness." Furthermore, "Freedom, manifesting itself through anguish, is characterized by a constantly renewed obligation to remake the self which designates the free being" (quoted after Cumming 1965).

Sartrian philosophy would seem propitious for suicide or for becoming a recluse. Far from that, Sartre has been intensely active, a member of the French Resistance during the Nazi occupation, then an immensely productive writer and a resourceful polemicist. His rancor is directed against self-righteous hypocrites and the entrenched bourgeois establishment. He sought a partnership with communism, but found the party orthodoxy too confining. He declared a keen-witted reprobate a "saint," because this wretch "holds the mirror up to us: we must look and see ourselves there."

Barzun (1964) credits the artists with having discovered "The dry rot of the soul" in the modern world, the blame for which this author lays squarely on science and *techne* (technology). Artists supposedly "detest and despise the scientific culture," because "Something pervasive that makes the difference, not between civilized man and the savage, not between man and the animals, but between man and the robot, grows numb, ossifies, falls away like black mortified flesh when techne assails the senses and science dominates the mind." The rebellion of the artist against society with its science and *techne* ends, however, in "treason." The modern artist becomes an ally of science in its work of "dehumanizing" which he rebelled against; he too discovers that the human condition, and life itself, are "hateful frauds."

Feelings of nausea, disgust, dry rot, emptiness, and of the hateful fraudulence of life, etc., are not exclusive privileges of the artistic and intellectual elites, of which Sartre and Barzun are eminent representatives. These feelings have percolated down to the "mass man." This is obviously not because Sartre's and Barzun's writings are favorite reading of the masses. The cause lies deeper than that. Certain experiences have been common to all social, economic, and educational strata. There were two world wars, in which the reputedly most highly civilized nations committed acts of unspeakable savagery. We know about Hitler's and Stalin's concentration camps and genocide programs. Brilliant scientific discoveries are subverted for military use. Material abundance and affluence turn to dust in our mouths, because so many feel alienated from the impersonal bigness which has produced this abundance. Many, especially among the young, are satiated with all the good things that ought to make life enjoyable—food, drink, sex, good clothes, money. When these good things come too easily, what one does not have to struggle for has little value and brings little satisfaction.

Probably everywhere and in every generation, there have been some people who felt that things were going badly in the world, or even from bad to worse. The term "alienation" is, however, of relatively recent vintage. At any rate, it is not a usual or a healthy situation for masses of people to feel alienated from the societies of which they are members. Jules Henry, a social anthropologist, finds this to be alarmingly general in the United States, and he entitles the book in which he presents the results of his studies *Culture against Man* (1963). "Ours," he says, "is a driven culture. It is driven by its achievement, competitive, profit, and mobility drives, and by the drives for security and a higher standard of living. Above all, it is driven by expansiveness." What undermines contentedness in this driven culture? As Henry sees it,

Except for professionals and executives most Americans are emotionally involved neither in their occupations (what they do) nor in their job (the place where they do it). What finally relates the average person to life, space, and people is his own personal, intimate economy: his family, house and car. . . . The fact that the majority have little or no involvement in the institutions for which they work means that work, which in most nonindustrial cultures of the world is a strong and continuous socializing agency is, in America, also desocializing. . . . To almost any American his working companions, however enjoyable, are inherently replaceable. The comradely group a man has on one job can be replaced by a similar one on the next. The feeling of being replaceable, that others can get along without one, that somebody else will be just as good, is an active depressant in the American character.

The grimmest chapters of Henry's analysis are those dealing with "human obsolescence," the fate of persons who are not so lucky as to die before the advent of the infirmities of old age. I believe it was Toynbee who remarked that the feature of our time, unprecedented in all known history, is that the aged no longer have an unquestioned claim to being taken care of by the fit members of their nuclear or extended families. In any case, Henry is "impressed with the gulf between the aged and the young, even when the aged are mentally alert; and this is because our culture is an avalanche of obsolescence hurling itself into the Sea of Nonexistence."

The feeling that life is meaningless, that the world is not going anywhere in particular, that man is an insignificant part in the scheme of things anyway, that history may end in a holocaust, and that whatever its end may be it is of little concern to those who live now and will soon be annihilated by death, has at least occasionally probably haunted everyone. Some people take refuge in traditional religions. Others escape into blatant hedonism. Still others, and probably a majority, take the easiest way out, and settle on acceptance, conformity, and modern versions of "fertility cults."

Life dedicated to building a home, raising a family, and then quietly passing away must have seemed meaningful enough to primitive man. Why should it not be good enough for everybody? In his interesting and provocative book, Gabor (1964) claims that such a life is satisfying to "the common man" but not to "the uncommon man." In Gabor's view,

For the common man life is a cycle. It starts with the discovery of the world around the child in which everything is new; it then goes on to the great discoveries of sex and love and culminates in the young family. It contains smaller cycles of work, rest, and recreation, of modest wishes and their fulfillment. It can be a happy life to the end if there is not much physical suffering and if the old man or woman has learned to love the new generation more than himself.

Gabor is optimist enough to think "that the material paradise of the common man cannot be far away." Paradise seems to be out of the reach of the uncommon man, who "wants to leave a world different from what he found; a better world, enriched by his personal creation. For this he is willing to sacrifice much or all of the happiness that the common man enjoys."

The dichotomy of common–uncommon man is a dangerous oversimplification. Its chief uses are to nurture personal and group conceits and arrogance. Carried to its logical conclusion it leads to the frightful doctrines of Dostoevsky's Grand Inquisitor, or of dictators like Hitler and Stalin and their would-be imitators. In reality, human natures form a spectrum ranging from the most common to the most uncommon man, as Gabor defines them. Moreover, many thousands, or even millions, of men who had at some times during their lives, chiefly in youth, ambitions to become "uncommon" failed for a variety of reasons to achieve their aspirations, and had to settle for whatever bits of common man's happiness happened to be within their reach. They may nevertheless harbor a hope that things will someday be different, and different for the better, from the world in which they live; moreover they hope against hope that their lives may somehow contribute a mite to the betterment, even if that be only indirectly.

After all, the common and the uncommon man are members of the same societies and civilizations, and this makes them heirs to similar historical experiences. Western civilization, as well as the great civilizations of the East, had the systems of the beliefs of their members molded by the higher religions. Higher religions point toward ideals of the "uncommon" man. This tends to make anything short of these ideals unacceptable as long-distance or ultimate goals, even for the "common" man. The reason is the same which makes reading matter designed for children and adolescents unsatisfactory as reading fare to adults who have become habituated to adult reading.

Suppose that you live for your children, that they will live for their children, and these will live again for theirs, and so on, ad infinitum. Biologically, this makes good sense; to some "common" as well as "uncommon" people this amounts only to a senseless repetition. We wish to see, or at least to hope for, something not only novel but also better. In the nineteenth century, most people in the West were convinced that progress does take place—more or less steadily or even unavoidably. We are no longer so sure. There is, however, a glimmer of hope for a new synthesis.

6

The Teilhardian
Synthesis

Among the two million or
more species now living on earth, man is the only one who
experiences the ultimate concern. Man needs a faith, a
hope, and a purpose to live by and to give meaning and
dignity to his existence. He finds himself in this world not
by his own choice; he wants at the very least to avoid ex-
cessive suffering, and to capture the joys that may be within
his reach. He desires to experience beauty and to shun ugli-
ness. Above all, he yearns for love and relatedness to other
persons; he wants to gain and to hold his self-respect, and
if possible the respect and admiration of others. For this
respect and self-respect, he may forgo pleasures and accept
pain and ordeal.

This is necessary, but is this sufficient to make life meaningful? To some people this is sufficient, *faute de mieux*.
But others, perhaps unreasonably, ask for more. If mankind is meaningless then my personal existence cannot be
meaningful. I must discover a hope for mankind in its historical development. The purpose of my life can only be
a small part of mankind's larger purpose. It is, furthermore, inconceivable for mankind to have meaning if the
universe has none. Man is involved in mankind, mankind
in life, life in the planet earth, and earth in the universe.
The universe of which mankind is a part must be meaningful. Toynbee (1956) has expressed this beautifully as follows:

The Human Spirit that dwells in each of us cannot refrain from seeking for an explanation of the Universe in which we find ourselves, and it insists that our Weltanschauung shall give the Universe significance without making the Universe center round the Self. In logic it may be impossible to reconcile these two requirements. Yet, even in the teeth of logic, the Human Spirit will not consent to abandon its search for explanation of the mystery; and the new gospel revealed by the higher religions does seem to offer a reconciliation in the intuition that the meaning of Life, Existence, and Reality is Love.

Fertility cults and the elemental joys of being alive sufficed to give at least a semblance of purpose to primitive peoples. Mankind's evolution has left this stage of happy spiritual childhood far in the past. Modern man must raise his sights above the simple biological joys of survival and procreation. He needs nothing less than a religious synthesis. This synthesis cannot be simply a revival of any one of the existing religions, and it need not be a new religion. The synthesis may be grounded in one of the world's great religions, or in all of them together. My upbringing and education make me biased in favor of Christianity as the framework of the synthesis. I can, however, understand people who would prefer a different framework. What is important is that the outcome must be truly a *synthesis*. It must include science, but it cannot be science alone, and in this sense it cannot be "scientific." It must include art and esthetics, but it cannot be esthetics alone. A faith which stands in flagrant contradiction with well-authenticated scientific findings cannot be right, but one in accord with such findings may nevertheless be wrong. Science discovers what exists; man has a longing to discover what ought to exist. The synthesis must be esthetically satisfying, but it must also be rationally persuasive.

The role of science in a religious synthesis has been stated with admirable clarity by Tillich (1963):

Of course, theology cannot rest on scientific theory. But it must relate its understanding of man to an understanding of universal nature, for man is a part of nature and statements about nature underlie every statement about him. . . . Even if the questions about the relation of man to nature and to the universe could be avoided by theologians, they would still be asked by people of every place and time—often with existential urgency and out of cognitive honesty. And the lack of answer can become a stumbling block for a man's whole religious life.

To satisfy man's hunger for meaning, not only man but the whole of nature, living and nonliving, must be understood in their relatedness. For man, though he may be nature's spiritual vanguard and spearhead, is nevertheless only a small part of nature. Viewed on a cosmic time scale, he is very much a newcomer in the universe. He appeared at most two million years ago, while life on earth appeared probably about two billion years ago, and the universe is five to ten billion years old. If man be regarded in some sense as a being above nature, it is certain that he has only recently emerged from nature, on the bosom of which he took shape.

Life is very much older than man, and the universe is much older than life. This points to an indispensable condition which any synthesis must satisfy in order to be acceptable. It must envisage man, life, and the universe as changing rather than fixed, as parts of a single ongoing process rather than as three separate static realms. The central postulate of the synthesis must be that the universe and everything in it are evolving products of evolution. The synthesis must be an evolutionary synthesis.

This is not easy for some people to accept. Man generally yearns for stability and distrusts change. Even though people see changes taking place all around them, they still like to believe that there is something basic, some general scheme of things, which remains forever stable and immutable. Stability suggests security, and change is liable to bring hazard, exertion, insecurity. It would seem that scientists, more than any other group of people, should not only be habituated to but should welcome change. Science is cumulative knowledge; Aristotle did not know many things which are now known to every schoolboy. We must hope that the scientists who succeed us will know more than we do, and will understand things better than we understand them. Scientists are, however, human, and some of them cannot help feeling a secret resentment when some of their ideas are shown to be invalid.

Religions, on the contrary, often pride themselves on being forever immutable. Religion is said to be the sheet anchor, a symbol of stability. And yet Toynbee (1956) writes: "What is permanent and universal has always and everywhere to be translated into terms of something temporary and local, in order to make it accessible to particular human beings here and now." An evolution of religion is therefore not incompatible with possession of permanent and universal verities. The seeming incompatibility arises because of the failure to distinguish between what is permanent and universal and what are merely historical accretions in religious teachings (see page 109). A parallel situation is found in the realms of art and of esthetics. We admire the drawings made by Paleolithic man in the caves in which he lived. The beauty which at least some individuals esteemed at the dawn of cultural evolution is esteemed by us today. This does not preclude an evolution of artistic and esthetic perception. The Altamira and Lascaux cave paintings have not made superfluous those of Michelangelo and Picasso, nor vice versa. Malraux (1960) finds that "all art forms embodying an aspect of the inapprehensible" have in common "their revelation of the presence of an Other World, not necessarily infernal or celestial, nor merely a world beyond the grave; rather a supra-real world, existing here and now. For all alike, in different degrees, the 'real' is mere appearance and something else exists, that is not appearance—and does not always bear the name of God."

Affirmation of the reality of time and of history is the crux of evolutionary doctrine. If evolution did not happen, then time is inconsequential and history is "a tale told by an idiot, full of sound and fury, signifying nothing." Those who nevertheless wish to believe that the world and human nature are essentially stable fall back on such intellectual contrivances as myths of eternal return and cyclic theories of history. The most ingenious contrivance of this kind, which has endured for more than two millennia, is Plato's theory of ideas. Things which we observe are only shadows of the perfect, unchangeable, eternal ideas; the shadows may shift to and fro, the eternal *eidos* is serenely stable.

We have said above that Christianity is basically evolutionistic. It affirms that the meaning of history lies in the progression from Creation, through Redemption, to the City of God. As Francoeur (1965) convincingly shows in his thoughtful book, the early Christian thinkers and Fathers of the Church did not always hold views which would at present be described as fundamentalist. The thinking of their successors was, however, influenced by the longings of their flocks for stability and security. They settled on a compromise. The Creation was assumed to have happened within a few days, some five thousand years ago, and nothing of much consequence is happening at present. The Creation has given us a cozy Earth, with the sun, moon, and stars going round and round for man's convenience and delectation, and with animals and plants provided to feed man and occasionally to scourge him. The changes that we can see are for the most part cyclic, exemplified and symbolized by the phases of the moon and by the annual succession of seasons, and the death and rebirth of vegetation. The City of God will arrive suddenly and unexpectedly. It will not be an achievement of man's historical efforts to move toward his Creator, but rather a spontaneous gift of God, as the Redemption was.

Religious conservatism goes easily together with social and political conservatisms. Innovations are to be distrusted; what was good for our fathers and grandfathers is good enough for us; ancient customs, laws, and constitutions are good because they are ancient; they are safe, and have been shown by experience to be workable. Generation should follow generation, perhaps trying to accumulate more wealth and to populate more land, but leaving the social fabric and the prescribed rules of behavior unaltered. Some people will sin frequently, and all will do so sometimes, but duly performed rites will bring them absolution. Other rites will bless the births, marriages, and deaths, and so it will go on forever.

The foundations of all conservatisms are undermined by the flood of scientific discovery. The first telling blow was the demonstration that the earth travels around the sun, and not vice versa. The world happens to be some billions, instead of a few thousand years old. These facts could conceivably be reconciled with and accommodated in the scheme of a static world and of immutable essences. More fateful has been the realization that the Creation is an ongoing process, not an event of a distant past. Change, evolution, is not an illusion and not a regrettable exception to the rule of stability; it is, on the contrary, an all-pervading mood of nature. The hostility to Darwin's theory of biological evolution arose not only because Darwin showed that man is a relative and a descendant of lowly animals. This hostility had also a deeper source. If the whole of nature and living species change and improve in the process, then no traditional order can plausibly be believed to be unalterable and sacrosanct. In an article written in 1920 but published almost forty years later, Teilhard de Chardin (English version in 1964) describes the intellectual ferment introduced by evolutionism as follows:

It is a pleasant and dramatic spectacle, that of Mankind divided to its very depths into two irrevocably opposed camps— one looking towards the horizon and proclaiming with all its new-found faith, "We are moving," and the other, without shifting its position, obstinately maintaining, "Nothing changes. We are not moving at all."

Ethical, ideological, and philosophical implications of evolution have been considered by many thinkers. Suffice it to mention Herbert Spencer, T. H. Huxley, Julian Huxley, and C. H. Waddington in England; John Dewey and G. G. Simpson in the United States; E. Haeckel, M. Hartmann and B. Rensch in Germany; and H. Bergson in France. It remained, however, for the inspired seer, P. Teilhard de Chardin, a French Jesuit and a paleontologist, to relate evolution to the ultimate concern, and to sketch a synthesis in which evolution is "a light illuminating all facts."

Teilhard's works have evoked enthusiastic praise as well as harsh, and even vitriolic, criticism. Scientists distrust theologians dabbling in science, just as theologians distrust scientists barging into theology. Teilhard was, however, both a scientist and a theologian. Some of his overzealous followers claim that his work is equal in importance to Darwin's. This can hardly be sustained, because the Teilhardian synthesis does not have the force of a scientific demonstration. Its intellectual grandeur may nevertheless be recognized even by those unconvinced of its validity. Teilhard's writings belong really to a class by themselves; an understanding of their singularity is essential for a comprehension of their contents.

Teilhard was a Christian mystic, who happened also to be a scientist, and who had in addition a gift of poetic imagery. He recognized that what he was writing about was his "fundamental vision" and "conviction strictly undemonstrable to science." And yet in his greatest work, *The Phenomenon of Man*, written between 1938 and 1940, published in French in 1955 and in English translation in 1959, he claimed that what he wrote was "purely and simply a scientific treatise." (Unless specified otherwise, all the Teilhard quotations given in the following pages are from this work.) The claim of being "scientific" made Teilhard easy prey for critics, who pointed out that his work deviates from the accepted style of scientific discourse. It is nevertheless unfair to describe Teilhard's views as "mystical Christianity ostensibly derived from evolutionary principles" (Simpson 1965). The idea that Christianity can be derived from evolutionary principles, or from any other scientific findings, would have seemed monstrous to Teilhard. What he tried to do was something entirely different, namely to create a coherent *Weltanschauung,* including his mystical Christianity as well as his scientific knowledge.

It would, then, be nearer the truth to say that Teilhard saw science illuminated by his mystical insights. Teilhard's achievement is best describable in the words of Toynbee (1956, written not in connection with Teilhard's ideas, which were then unknown to Toynbee):

The Truth apprehended by the Subconscious Psyche finds natural expression in Poetry; the Truth apprehended by the Intellect finds its natural expression in Science. . . . On the poetic level of the Subconscious Psyche, the comprehensive vision is Prophecy; on the scientific level of the Intellect it is Metaphysics.

Teilhard was a prophet and a metaphysician in the special sense in which Toynbee used these words.

Some people feel that science and poetry mix no better than oil and water. They may as well spare themselves the effort of reading Teilhard's works. Teilhard has addressed himself to those unwilling to tolerate ideological schizophrenia. Those who are looking for an esthetically as well as rationally satisfying synthesis, instead of an intellectual life divided into isolated compartments, can find in Teilhard a help and an inspiration. It is not my intention to review here the whole range of Teilhard's ideas; it is rather to scrutinize his synthesis from the standpoint of modern biology, and perhaps to suggest some modifications. If my remarks seem to some of Teilhard's admirers to be too often critical, I can only say that nothing could be more damaging to his synthesis than to have it frozen in a fixed canon. Teilhard had his biological views formed mostly in the nineteen-twenties, and much has happened in biology since then. If his biology is antiquated, that of Darwin is still more so. Nor should it be forgotten that any synthesis can have only a temporary utility, because at least some of the components of any imaginable synthesis will themselves be undergoing changes. Science and art and philosophy will, it is to be hoped, be different in the future from what they are today. Every generation will face the task of revising and renewing the synthesis.

"Men's minds are reluctant to recognize that evolution has a precise *orientation* and a privileged *axis*." This is the cardinal postulate of the Teilhardian synthesis. Evolution, human and biological and cosmic, is not simply a lot of whirl and flutter going nowhere in particular. It is, at least in its general trend, progressive. The evolution of life is a prolongation of the evolution of nonliving matter; human evolution is an extension of biological evolution; and the "megasynthesis" which Teilhard prognosticates will be a sequel to human evolution. Man's individual life is a component part of the evolution of the universe; man's ultimate concern, and his individual meaning and dignity are atoms of the meaning of the whole cosmos.

Some writers restrict the word "evolution" to biological evolution only. This seems to me gratuitous. The universe has had a historical development; so had life, and so had mankind. This historical development did advance to life from absence of life, and did ascend to man from nonhuman ancestors. Although inorganic evolution is due to operation of agencies different from the organic, and human evolution has again causes of its own, life is newer than the universe, and man is newer than life. As shown in Chapter 3, the origin of life and the origin of man may be regarded as evolutionary transcendences which opened up possibilities for developments of new kinds. Viewed in the perspective of time, these transcendences do represent breaks in the evolutionary continuity, but not events unprepared by the foregoing developments. They show that evolution as a whole has been progressive, if the word "progress" is to mean anything.

Teilhard's assertion that the evolutionary process has a definite orientation must be very carefully examined. Evolutionary changes taking place at any given time are conditioned by the changes which preceded them, and they will condition the changes that take place in the future. This is especially obvious in biological evolution—the evolutionary past of a living species is, as it were, inscribed in its genes. Evolution is not a collection of independent and unrelated happenings; it is a system of interrelated events. Life could not have arisen until cosmic evolution had produced at least one planet capable of supporting life. A being such as man, with a capacity for symbolic thinking and for self-awareness, could not have appeared until biological evolution had generated organisms with highly developed brains. Since certain evolutionary events could have happened only on the foundation of a series of preceding events, the history of the universe may be said to have an "orientation." We may choose to call the evolutionary line that produced man the "privileged axis" of the evolutionary process.

"Orientation" may, however, be understood also in a different manner. The process of evolution may be oriented, guided, and propelled by some natural or supernatural agency. Evolution was then able to follow only a single path, so that its final outcome, as well as all the stages through which it had to pass, were predestined and have appeared in a certain fixed order in time. Some minority schools among biologists believe in such a foreordained orientation. For example, the finalists posit that all evolution occurred for the express purpose of producing man, and that evolutionary changes at all times were guided toward this goal by some supernatural power or powers. Another school maintains that evolution, at least biological evolution, is orthogenesis. The changes that occur are sequences of events determined by factors inside the organism, by the structure of its genetic endowment, and proceed straight toward a fixed objective, such as man. The evolutionary development follows, then, a predetermined path, and its final outcome is likewise predetermined.

In contradistinction to finalism, orthogenesis does not necessarily assume supernatural forces guiding evolution. The favorite "explanation," which is really nothing more than an attractive analogy, is that evolutionary development (phylogeny) is predetermined in the same way as is the development of an individual (ontogeny). A fertilized human egg cell does not contain a homunculus, a little human figure, and yet from this egg cell arises an embryo, which undergoes many complex transformations and growth and finally becomes an adult man. Did the bodies of our remotest ancestors, or even primordial life, contain all the rudiments needed to produce all evolutionary developments? If evolution is orthogenesis, then it is what the etymology of the word "evolution" implies, i.e., unfoldment of preexisting rudiments, like the development of a flower from a bud. Finalism and orthogenesis have this much in common: the evolutionary history of the living world was predestined at the beginning of life and even in primordial matter. If true, this would make evolution a rather dull affair. Evolution produced nothing really new, since all that it did produce was ordained to happen. It lacked all freedom and creativity.

Teilhard declared it to be his "considered opinion" that orthogenesis is "essential and indispensable." The meaning which he ascribed to this term was, however, an unusual one. He gave two definitions on the same page (page 108 of the English version of *The Phenomenon of Man*). One of them is: "a law of controlled complication, the mature stage of the process in which we get first the micro-molecule then the mega-molecule and finally the first cells." The other is: "The manifest property of living matter to form a system in which terms succeed each other experimentally, following the constantly increasing values of centro-complexity." What this seems to mean is a statement of the undoubted fact that, seen in retrospect and in its totality, evolution was indeed progressive, and in this sense directional and oriented.

The evidence of progress and directionality in biological evolution is clear enough if the living world is considered as a whole. To be sure, in some groups, such as various parasitic forms, evolution was often retrogressive (loss of organs, particularly degeneration of the nervous system); in other groups evolution was seemingly given to production of endless variations on the same theme. And yet, the net outcome of evolution is that today the earth is no longer populated exclusively by primordial viruses and amoebae. Numerous complex organisms, with body structures that can only be compared to works of art, have appeared. Most remarkably, evolution has produced organisms with highly developed nervous systems, which convey to them information about the states of their environments. To some extent, such organisms can dominate their environments, instead of being dominated by the latter.

Now, it is the totality of evolution that occupies Teilhard's attention almost exclusively. The only particular evolutionary line which interests him is that of man, and this because he believes that in man evolution as a whole is, as it were, brought into focus. He even envisages the entire living world as being a single organism: "Taken in its totality, the living substance spread over the earth—from the very first stages of evolution—traces the outlines of one single and gigantic organism." This is a very perceptive, and beautiful, metaphorical statement of the unity of life. The evolution of the symbolic supraorganism is, however, a problem separate from that of the evolution, or better the evolutions, of the millions of particular species of living beings which inhabit our planet at present and inhabited it in the past. Moreover, orthogenesis is a hypothesis which endeavors to explain what causes evolution, rather than a summary description of the evolutionary history of the living world.

This puts me, willy-nilly, in the peculiar position of having to argue that, in spite of himself, Teilhard was not an exponent of orthogenesis. Concerning the causes of evolution he had actually little to say. Although the latter part of his life coincided with the development of the modern biological (synthetic) theory of evolution, he had only a hazy idea about it. And yet his general conception of the nature of evolution harmonizes with the fundamentals of biological theory far better than with that of orthogenesis. And let this be made clear: what is here involved is not a technical biological problem; the issue is critical for the whole Teilhardian synthesis.

If evolution follows a path which is predestined (orthogenesis), or if it is propelled and guided toward some goal by divine interventions (finalism), then its meaning becomes a tantalizing, and even distressing, puzzle. If the universe was designed to advance toward some state of absolute beauty and goodness, the design was incredibly faulty. Why, indeed, should many billions of years be needed to achieve the consummation? The universe could have been created in the state of perfection. Why so many false starts, extinctions, disasters, misery, anguish, and finally the greatest of evils—death? The God of love and mercy could not have planned all this. Any doctrine which regards evolution as predetermined or guided collides head-on with the ineluctable fact of the existence of evil.

Philosophers have struggled with the problem of evil for more than two millennia. Teilhard certainly knew all this, and knew that the only hope for a solution lies in the replacement of predestination by freedom as the mainspring of creation. On the human level, freedom necessarily entails the ability to do evil as well as good. If we can do only the good, or act in only one way, we are not free. We are slaves of necessity. The evolution of the universe must be conceived as having been in some sense a struggle for a gradual emergence of freedom. The outcome of evolution is not predestined because, in Teilhard's words, "There is a danger that the elements of the world should refuse to serve the world—because they think; or more precisely that the world should refuse itself from perceiving itself through reflection." Here Teilhard's ideas draw near those of many other thinkers, such as Hartshorne (1962), a philosopher who is at least aware of evolutionary problems.

Teilhard describes the method of evolution as groping ("tâtonnement"). This is a more poetic and impressionistic than a rigorously scientific characterization, and yet it is remarkably apposite. Some of the fundamental tenets of modern biological theory have been outlined briefly in Chapter 3; here it is necessary to consider some of them further from the standpoint of their bearing on the problem of predetermination vs. creativity of biological evolution. Is evolution merely a painfully, and seemingly needlessly, slow maturation and unfoldment of what was always there, preformed and foreordained to be unveiled in the course of time? Or can it be a succession of trials and errors, some of them resulting in inventions? Can evolution be the realization of just a tiny fraction of an infinite series of potentialities? And can it be understood as in any sense a response of the creation to its Creator in increasing freedom? Are the pain and struggle and evil connected with this gradual expansion of freedom?

Evolution has three stages or levels: (1) production of genetic raw materials through mutation, (2) formation through natural selection and Mendelian recombination of genetic endowments adapted to survive and reproduce in certain environments, and (3) establishment of species barriers by reproductive isolation. Mutations are changes in the genes and chromosomes. Their outstanding property is adaptive ambiguity. This means that a mutation arises regardless of whether it may be useful or harmful to the organism. In point of fact, most mutations are harmful, many produce hereditary defects or diseases, and some are lethal. How, then, is it possible that mutation supplies the genetic raw materials of evolution? The answer is that a minority of mutations are not harmful but useful, especially when the environment in which a biological species lives is altered. Some of the old genes then become no longer adaptive, and some mutant genes replace them.

The harmfulness of most mutation is a dramatic demonstration of the absence of guidance in evolution. At the level of mutation, evolution is neither directional nor oriented nor progressive. It is the very antithesis of orthogenesis. Mutation alone would cause chaos, not evolution. Natural selection redresses the balance. Harmful genes are reduced in frequency, and useful ones perpetuated and multiplied. As pointed out above, some authors liked to compare natural selection with a sieve, which retains some particles and lets others go through. This is too crude an analogy. Natural selection works not with genes but with whole genetic endowments; what survives or dies, begets progeny or remains childless, is not a gene but a living individual. A gene useful in combination with some genes may be harmful in combination with others. The changes which natural selection promotes at present depend upon the changes that occurred in the past. Natural selection is comparable not to a sieve but to a regulatory mechanism in a cybernetic system. The genetic endowment of a living species receives and accumulates information about the challenges of the environments in which the species lives. The evolutionary changes are creative responses to the challenges of the environment. They are not alterations imposed by the environment as Lamarckists mistakenly thought.

Almost all higher organisms and many lower ones reproduce sexually. Sex is a supremely efficient method to generate countless new genetic endowments, which are exposed to the arbitrament of natural selection. This is a corollary of Mendel's laws; in a population which has n genes each represented by two variants, 3 to the nth power of different genetic endowments are possible. If n is of the order of hundreds or thousands, the number of potentially possible gene combinations far exceeds the number of individuals of any species existing on earth. In a sense, the Mendelian mechanism is, then, more efficient than it needs to be. A vast majority of potentially possible genetic endowments will never be realized.

The consequences of this prodigious efficiency are nevertheless very interesting. First, every individual in a sexually reproducing species, such as man, has a genetic endowment which is unique, unprecedented, and nonrecurrent (identical twins are, however, an exception, being genetically identical or very nearly so). Secondly, evolutionary changes are unique events; for example, the evolution of man from his prehuman ancestors is infinitely unlikely to be either repeated or reversed. If life exists on some planets other than our earth, it is utterly improbable that the evolutionary process there went exactly as it did here. Thirdly, among the myriads of the gene combinations that could be formed, some would be adaptively harmonious and others disharmonious. How to maximize the frequency of the former and to minimize the latter? Life has evolved different organisms adapted to different environments and different ways of life. Mixing their genes would almost always be disadvantageous. There are at least two million (some estimates double this figure) biological species on earth. The species do not interbreed and do not exchange genes, or do so rarely. They are reproductively isolated or nearly so. The reproductive isolation is accomplished in a great variety of ways. The species may breed at different seasons, may occur in different habitats, the sexual attraction between them may be weak or absent, the hybrids if produced may be inviable, weak, or sterile, etc.

Two kinds of evolutionary changes can be distinguished. First, there is anagenesis: as its environment changes with time, a biological species undergoes changes that maintain or improve its adaptedness, but continues to be a single species. The second is cladogenesis, splitting up of a single species into two or more derived ones. This happens most

easily when members of a species live in different terri-
tories with different environments. They become differen-
tiated first into geographic races, which natural selection
makes genetically more and more different in response to
their respective environments. When the genetic divergence has gone so far that the gene exchange becomes
adaptively disadvantageous, natural selection promotes the
development of reproductive isolation, and consequently
the breaking up of an ancestral species into two or several
new ones. Both anagenesis and cladogenesis occurred in the
evolution of most animal and plant groups. In human
evolution there has been a predominance of anagenesis. To
be sure, the human species is, and presumably always was,
composed of several races. Races differentiate, diverge, or
fuse and disappear; at some periods divergence, and at
others fusion, predominates. Race differentiation is, of
course, cladogenesis. However, at least since the middle of
the Pleistocene period there was always only a single human
species. *Homo erectus* with its several races evolved into
Homo sapiens, modern man. The races of *Homo sapiens*
show no tendency to diverge further; in point of fact, for
at least two millennia they have been converging, and the
convergence is gathering speed.

The process of organic evolution can be epitomized also
in the terms used by Toynbee in his monumental *Study of
History* to describe the origins of civilizations and their
subsequent histories: challenge and response. Challenges
come from the environment, the responses occur through
the agency of natural selection. In the absence of challenge
there is evolutionary stagnation; hence the "living fossils,"
species that show no perceptible change over long stretches
of geological time. But a challenge does not guarantee that
a response will be given, nor does it determine exactly what
the response will be. A species which fails to respond because of the absence of suitable genetic materials will go
into decline or become extinct. A response may be the
transformation of the species into a new state (anagenesis),

or splitting into several species, adaptively specialized for different environments (cladogenesis). Cladogenesis enhances organic diversity; it may be said to create numerous trial parties "exploring" the possibilities of different environments ("groping" in Teilhard's more picturesque simile).

The evolution of life took a long time, two billion years at least, because it was not a simple matter of unfolding or uncovering something which was there from the beginning. The history of life is comparable to human history in that both involve creation of novelty. Both proceed by groping, trial and error, many false starts, being lost in blind alleys, failures ending in extinction. Both had, however, also their successes, master strokes, and both achieved an overall progress. There has been a general acceleration of evolution. Cosmic evolution took more time than the biological, and human evolution is of shortest duration. The concepts of creativity and freedom are not directly applicable below the human level. It may nevertheless be argued that the rigid determinism is becoming gradually relaxed as the evolution of life progresses. The elements of creativity are more perceptible in the evolution of higher than in that of lower organisms (I have discussed this matter in more detail elsewhere—Dobzhansky 1963).

Adherents of finalism and orthogenesis contend that since it is quite incredible that evolution could all be due to "chance," one must assume that it has had a design which it has followed. The reality is, however, more complex and more interesting than the chance vs. design dichotomy suggests. The statement that life must have had from the very beginning the potentialities for all the evolutionary developments which did in fact occur is obviously true but just as obviously trivial. If it were otherwise evolution could not have done what it did. What is more important is that life had also innumerable other potentialities which remained unrealized. This follows from the fact that only a minuscule fraction of the possible gene combinations can ever be actualized. It is misleading to think that all organisms, including man, were preformed in the primordial life, and merely needed time to unfold, like flowers from buds. The real problem lies in a different dimension.

A child receives one-half of the genes of his father, and one-half of the maternal ones; which particular maternal and paternal genes are transmitted to a given child is a matter of chance. Which mutations occur, and when and where, is also a matter of chance. And yet, evolution is a matter of chance no more than a mosaic picture is a fortuitous aggregation of stones. (Some modern "painters" make "pictures" by pelting a canvas with paint, but there seems to be no biological analogy to this stunt.) In evolution, chance is bridled in by an antichance agency, which is natural selection responding to environmental challenges. Let us be reminded that natural selection does not act as a sieve, but as a much more sophisticated regulating and guiding device. What is most remarkable is that the "guidance" does not amount to a rigid determinism. Especially in the evolution of higher organisms there are discernible elements of creativity and freedom.

How is this possible? Many environmental challenges can be met successfully in more than one way. For example, all desert plants must cope with dryness. Different plants do so, however, by different means. Some have leaves reduced to spines, others have leaves protected by waxy or resinous secretions, others shed their leaves when humidity becomes deficient, and still others germinate, grow, flower, and mature seeds all within a short span of time when water is available. Animals above a certain body size must have some organs of respiration; these may be gills, or tracheae, or lungs of a variety of kinds. Several groups of animals have evolved flight, and again did so in different ways. To live in cold countries, animals may have warm fur and a metabolism which maintains a constant body temperature, or they may be dormant during the cold season, or they may migrate to warmer countries. Some animals avoid their enemies by concealing shapes and colorations, which make them hard to see; others have gaudy warning colorations, which advertise their presence and their real or pretended dangerous or obnoxious qualities to their potential enemies. All these methods of adaptation are sometimes successful, and none seems intrinsically superior to the others.

The multiplicity of ways of becoming adapted to similar environments is not in accord with hypotheses of design and orthogenesis in evolution; these hypotheses would lead one rather to expect that a single, and presumably most perfect method, will be used everywhere. On the contrary, natural selection is more permissive. Nineteenth-century evolutionists called natural selection "the survival of the fittest"; we prefer to delete the superlative: what is fit to survive survives. One of Teilhard's aphorisms is that evolution is "pervading everything so as to try everything, and trying everything so as to find everything." This is an apt characterization, but it needs a correction. Teilhard was unaware of the fact which we stressed above, that only a minuscule fraction of the potentially possible gene combinations are ever actualized. "Trying everything" is an overstatement; evolution has tried and found far from "everything" that could be found.

The gene combinations which now compose the genetic endowment of the human species did not exist in, say, Eocene or Cretaceous time. Could a biologist, if one had lived in these remote times, have predicted that the human species would eventually evolve? This is not as difficult a question as it might seem—if the ancient biologist had only a knowledge comparable to our present one, he could not have made such a prediction. Let it be noted parenthetically that a question almost equally fanciful is at present debated quite earnestly by some popular writers and by governmental agencies with handsome budgets: If there is life in many parts of the universe outside of our little planet, have manlike creatures also evolved in many places? I wish to plead that the headstrong assertions that such creatures must have evolved are premature, to say the least. Living on an insignificant little planet, not too unlike billions of other lifeless ones, we may well be the only rational beings in the universe.

As mentioned above, Teilhard describes the method of evolution as "groping." He also claims that "Groping is directed chance." This requires careful examination. Among Teilhard's many metaphors, "groping" is perhaps the most ingenious one. Natural selection operates with mutations and gene combinations in the origin of which "chance" plays an important role. Natural selection "directs" this "chance" into adaptive channels. One must, however, beware of personalizing natural selection. It is not some kind of spirit or demon who directs evolution to accomplish some set purpose. "Groping" in the dark is, indeed, the only way natural selection can proceed. Now, groping may lead to discovery of openings toward new opportunities for living. It may also end in a fall from a precipice. It may preserve and enhance life, or it may lead to extinction. Teilhard was a paleontologist, and he was quite familiar with extinction of evolutionary lines. Yet he devoted strangely little attention to this phenomenon in his writings. It would have caused him no difficulty had he realized that natural selection is necessarily opportunistic and shortsighted in its gropings. Lacking a prevision of the future, natural selection adapts the living species to the environments which exist here and now. A species may be well-adapted to the present environments, but when the environments change it may be unable to readapt quickly enough, and may become extinct. High adaptedness does not always go together with high adaptability.

Taken literally, "directed chance" is a paradox. Teilhard adds to it still another paradox: in evolution there "are so curiously combined the blind phantasy of large numbers and the precise orientation of an end pursued." Yet these paradoxes give a trenchant description of what actually happens. How is this possible? Biological evolution has been in part cladogenesis (see above), and the number of species living on earth has been increasing with time. The species living at present are descended, however, from only a minority of the species of the past. The minority becomes smaller and smaller as we go farther back in time. This is because a majority of the evolutionary lines end by becoming extinct. A minority do remain living, and an even smaller minority achieve novel forms of adaptedness by means of new ways of life in new, previously unexploited or inefficiently exploited environments. The total organic diversity is gradually increasing, owing to the evolutionary branching, cladogenesis. Evolution has achieved more than to preserve life on earth from destruction. It has created progressively more complex and adaptively more secure organizations. The human species has attained the peak of biological security. It is unlikely to become extinct because of any conflicts with its physical or biological environments. Man is able, or soon will be able, to control his environments successfully. Extinction of mankind could occur only through some suicidal madness, such as an atomic war, or through a cosmic catastrophe.

The diversity of living beings on earth is fascinating, and at times a bit puzzling, to a biologist. Why, indeed, should there be more than two million species of organisms, some of them seemingly strange, almost whimsical creatures? The answer is perhaps that this "blind phantasy of large numbers" has made possible the evolutionary progress. The chief characteristic, or at any rate one of the characteristics, of progressive evolution, is its open-endedness. Conquest of new environments and acquisition of new ways of life create opportunities for further evolutionary developments.

The "discovery" of respiration by lungs permitted the descendants of a certain kind of fishes to emerge on land, to exploit its food resources, and to evolve into amphibians. Most amphibians continue, however, to be dependent on water for their developmental stages (tadpoles). This dependence was cast off by the "invention" of eggs protected by shells, inside of which the embryonic development is completed. The family of amphibians which in its "gropings" hit upon this "invention" evolved into reptiles. Some of the reptiles live in driest deserts, and many of them take no liquid water at all, deriving their entire water supply from the food they eat. The next key acquisition was probably that of the physiological mechanisms which maintain a constant body temperature regardless of the temperature of the environment. This may have happened independently in two groups of reptiles, which gave rise respectively to birds and to mammals. Birds developed flight and the ways of life made possible by flight. Mammals evolved a complex physiological machinery for the development of their embryos inside the mother's body, for feeding the infants on milk, and for parental care. These innovations were accompanied by increases in the brain size and by growth of mental abilities.

The magnitude of the evolutionary change in mental abilities between a fish and a mammal is impressive. Yet it is dwarfed by the restructuralization of the mental faculties which took place during the last million or two million years, geologically a short time, in the evolutionary line culminating in a single species, mankind. This change was great enough to deserve the name of transcendence. It is comparable only to that much more ancient transcendence, the origin of life. A psychological abyss seems to separate *Homo sapiens* from all other animals. In Teilhard's words, "Admittedly the animal knows. But it cannot know that it knows—this is quite certain." Man is able "no longer merely to know, but to know oneself; no longer merely to know, but know that one knows." However, let it be emphasized once more (cf. Chapter 3) that every basic human mental faculty is found in a rudimentary form in some animals. The differences between man and animals are, then, merely quantitative. This does not entangle us in a contradiction. Quantitative differences may grow large enough to become qualitative. Our animal ancestors did possess paltry beginnings of the faculties which, greatly developed and combined together, gave an altogether new kind of being—man.

Prior to the origin of life, the crust of the earth (lithosphere) was covered with a watery (hydrosphere) and a gaseous (atmosphere) envelope. Life has intercalated a new envelope, the biosphere, mainly at the interphase of the atmosphere with the lithosphere and the hydrosphere. Mankind has spread all over the globe, and become the newest, the "thinking envelope." For this, in the fullness of time probably the most important envelope, Teilhard and Vernadsky have suggested, apparently independently, the same name, the "noosphere." In at least two senses, the evolution of the noosphere is a prolongation of that of the biosphere. First, the noosphere has emerged from the biosphere, as this latter had emerged earlier from the hydrosphere. Secondly, the noosphere is connected with the other envelopes by feedback relationships. Man's power as a biological and even a geological agent is constantly growing. In return, man is influenced by his environment. And yet, the evolution of the noosphere, the noogenesis, is basically a novel process.

Reference has already been made to Teilhard's idea that the totality of the living substance on earth constitutes a supraorganism. Mankind evidently was a part, or if you wish an organ, of this supraorganism. Teilhard expects that it will become in its turn a new unit of a different sort— "an organic superaggregation of souls." Tillich (1963) argues that human history is a process basically distinct from evolutionary "history" in general. The distinctive characteristics of human history are: "to be connected with purpose, to be influenced by freedom, to create the new in terms of meaning, to be significant in a universal, particular and teleological sense." I would only add a reservation, namely that, as with the basic human mental faculties, these distinctive marks of human history may be present in rudimentary form on the biological level as well.

"The blind phantasy of large numbers" has played an important role in the entire evolutionary process—from cosmogenesis, through biogenesis, to noogenesis. Cosmic evolution gave rise to multitudes of galaxies, suns, and planets. Life, and eventually consciousness and self-awareness, appeared on at least one of the myriads of celestial bodies. Large numbers are involved also in human evolution. Mankind is multitudinous and is growing more so. Are the multitudes supererogatory? They may seem so, in view of the fact that the intellectual and spiritual advances are chiefly the works of elite minorities. To a large extent, they are due to an even smaller minority of individuals of genius. The destiny of a vast majority of humans is death and oblivion. Does this majority play any role in the evolutionary advancement of humanity?

It may well be doubted whether the elites could exist by themselves without the nameless multitudes. The function of the multitudes is not limited, however, to serving as manure in the soil in which are to grow the gorgeous flowers of the elite culture. Only a small fraction of those who try to scale the heights of human achievement arrive anywhere close to the summit. It is imperative that there be a multitude of climbers. Otherwise the summit may not be reached by anybody. The individually lost and forgotten multitudes have not lived in vain, provided that they, too, made the efforts to climb.

The thesis that all human beings are partners engaged in "the common enterprise" was argued forcefully by the eccentric but original Russian thinker Fedorov (1832–1903; see about him in Zenkovsky 1953). Unaware of this predecessor, Teilhard develops similar arguments in an evolutionary context: "No evolutionary future awaits man except in association with all other men." Reluctant to admit it, Teilhard enters here the realm of prophecy. Yet he does not allow his prophetic vision to soar out of sight of the solid ground of cumulative knowledge. His prophecy is not scientifically provable; if it were so it would be prediction rather than prophecy. However, a prophecy may be compatible with, or contradictory to, scientific knowledge. This gives us a warrant to examine Teilhard's prophecy from the point of view of evolutionary biology, which is, after all, the same point from which Teilhard himself takes his departure.

The trend prevailing in the evolution of the noosphere, the noogenesis, is toward "planetization" and the "megasynthesis." This implies a radical convergence and integration of the physical, cultural, and ideological branches of mankind. Branching, cladogenesis, has played a subordinate but not unimportant role in human evolution for the last million or two years. It has created racial, national, social-class, and cultural divisions. Like the diversity on the biological level, human diversity served to "try everything so as to find everything." The other side of the coin is not pretty; differences among men have often inflamed hatreds, cruelty, strife, war (hot and cold), genocide, concentration camps. Social Darwinists, as un-Darwinian as they are antisocial, contend that strife and all its grim consequences are merely the wages which mankind has to pay for progress. Some biologists still sing paeans "in praise of waste."

Teilhard rejects social Darwinism. In noogenesis, the most powerful impetus toward progress comes not from strife or waste but from love. Replacement of strife by love already began in biological evolution, biogenesis. The classics of evolutionism described natural selection as a consequence of the struggle for existence. The "struggle" does not, however, always mean strife. Our modern view of natural selection sees it promoted by cooperation as well as by competition. Moreover, the importance of cooperation relative to competition has been growing as biological evolution has advanced. By and large, it is greater among higher than among lower animals. Teilhard makes love a basic agent of all evolution: "Driven by the forces of love, the fragments of the world seek each other so that the world may come to being. This is no metaphor; and it is much more than poetry." Well, it is a metaphor, it is poetry, and as such it manages to convey an important insight.

To make a paradox in the Teilhardian style, let us say that in progressive evolution we find a competition for co-operativeness. There is also an evolution of love; love ascends from sexual love, to brotherly love, to love of mankind, to love of God. Love unites without casting off the diversity. On the human level it is the means whereby a person as well as the species achieves self-transcendence. The megasynthesis is "a gigantic psycho-biological operation" in which love is the main agent, and which leads to the unity in diversity. As early as in 1920 (published in Teilhard 1964), Teilhard wrote:

It is Mankind as a whole, collective humanity, which is called upon to perform the definitive act whereby the total force of terrestrial evolution will be released and flourish; an act in which the full consciousness of each individual man will be sustained by that of every other man, not only the living but the dead.

The "planetization" of mankind is, in Teilhard's view, made inevitable by the swiftly increasing facility of communication and by increasing knowledge. Mankind inhabits the surface of only one rather small planet. Unless means are found to emigrate and to colonize other planets, people will finally have to learn to live harmoniously or at least peacefully with more and more numerous neighbors. The main point here is not only that population densities have grown and are growing, but even more that technological inventions facilitate travel and make possible almost instantaneous transmission of information and ideas to every corner of the world. Knowledge promotes spiritual growth as well as unification—"to be more is in the first place to know more."

Teilhard's prophecies of eventual planetization and megasynthesis may seem to be daydreams of a visionary, taking no account of the forces of evil and of the darker sides of so many human natures. Teilhard realized that his ideas are liable to be misconstrued as advocating a reduction of mankind to a state of vapid uniformity for some benign stereotype. Nothing was more alien to his thought. He faced the formidable problem of how to reconcile unanimity and megasynthesis with individuality, freedom, and what Brinton (1953) so aptly called "multanimity." It is best to quote his own words:

The Earth not only covering itself with myriads of grains of thought, but enclosing itself in a single thinking envelope, to become functionally a single vast Grain of Thought on a planetary scale. The multitude of individual reflections grouped and mutually reinforced in the act of one single unanimous Reflection.

Far from becoming all alike, or undergoing amalgamation or coalescence, as the noogenesis approaches the consummation of megasynthesis the human personalities are expected to grow in depth and to maximize their individual uniqueness. The meaning of an individual life is its inclusion in the evolutionary upswing of noogenesis. Even on the animal level, individuals are not interchangeable, because they are neither genetically nor developmentally identical. Noogenesis leads to affirmation, not to leveling of individuality.

Is there anything more than Teilhard's burning faith to bear out the bright hope of the megasynthesis? Can one rule out the polar opposite: disunion, dispersion, and arrogant self-assertion of the individual against mankind? The antithesis to megasynthesis is the ideal of the Dostoevskian Grand Inquisitor and the Nietzschean Superman. Without specifically mentioning Dostoevsky or Nietzsche, Teilhard recognizes the danger. Human freedom enables man to choose also a direction away from megasynthesis. Mankind may become a dust of independent and dissociated sparks of consciousness. Some of those sparks, being stronger, brighter, or perhaps simply luckier, than the rest, will "eventually find the road always sought by the Consciousness towards its consummation." Teilhard rejects this possibility as leading into an evolutionary blind alley. Spiritually matured mankind should be able to extricate itself from such a blind alley, because man is the only form of life which need not accept the direction of the evolutionary forces acting upon him, but can direct his evolution. An isolated individual is biologically as well as humanly an anomaly. Hermits and anchorites lived in vain, unless they somehow communicated their insights to other people, provided, of course, that they had insights worth communicating.

Self-assertion which makes an individual break away from humanity is inimical to the growth of the person as well as of humanity. Healthy growth is fostered by love and arrested or stunted by egocentrism or egoism. Self-fulfillment is possible only through love for, and in a spiritual union with, others. "The true ego grows in inverse proportion to egoism." Teilhard parts company with the Buddhist ideal of blending and dissolution of the human personality in a union and eventual fusion with the Deity. Teilhard rejects such a dissolution. Man must "seek to reconcile the hopes, indispensable to him, of an unlimited future with the perspectives of his individual death which is inevitable."

The eventual consummation of all evolution is envisaged by Teilhard as a convergence in the Omega. "I am the Alpha and the Omega, the first and the last, the beginning and the end" (Rev. 22 : 13). This Christian religious symbol is beyond doubt the source and the inspiration of Teilhard's vision. He makes this unmistakably clear in his "Turmoil or Genesis?" (English version, 1964):

To the Christian, for whom the whole process of hominisation is merely the paving of the way for the ultimate Parousia, it is above all Christ who invests Himself with the whole reality of the Universe; but at the same time it is the Universe which is illuminated with all the warmth and immortality of Christ.

It is evidently the inspiration of a mystic, not a process of inference from scientific data, that lifts Teilhard to the heights of his eschatological vision. Yet he remains a consistent evolutionist throughout. The point which he stresses again and again is that man is not to be a passive witness but a participant in the evolutionary process.

The consummation of the World, the gates of the Future, the entrance into the Superhuman, they do not open either to a few privileged or to one chosen people among all peoples! They will admit only an advance of *all together,* in a direction in which all together could join and achieve fulfillment in a spiritual renovation of the Earth.

Bibliography

Where dates of publication of American and British editions differ, the date which appears in the text refers to the American edition.

Barzun, J. *Science, the Glorious Entertainment.* New York, Harper & Row; London, Secker & Warburg, 1964.

Bernal, J. D. "Molecular Matrices for Living Systems." In S. W. Fox (Ed.) *The Origin of Prebiological Systems.* New York, Academic Press, 1965.

Bidney, D. *Theoretical Anthropology.* New York, Columbia University Press; London, Oxford University Press, 1953.

Birch, L. C. "Creation and the Creator." *Journal of Religion,* 37 : 85-98, 1957.

—— *Nature and God.* London, SCM Press, 1965.

Blum, H. F. "Dimensions and Probability of Life." *Nature,* 206 : 131-2, 1965.

Brinton, C. *The Shaping of the Modern Mind.* New York, New American Library, 1953.

Brunner E. *The Christian Doctrine of Creation and Redemption.* New York, Westminster Press; London, Lutterworth, 1952.

Bury, J. B. *The Idea of Progress.* New York, Macmillan, 1932; London, Macmillan, 1920.

Calvin, M., and Calvin, G. J. "Atom to Adam." *American Scientist,* 52 : 163-86, 1964.

Cassirer, E. *An Essay on Man.* Garden City, Doubleday Anchor, 1944.

Choron, J. *Death and Western Thought.* New York and London, Collier, 1963.

—— *Modern Man and Mortality.* New York, Macmillan, 1964.

Coulson, C. H. *Science and Christian Belief.* New York, Oxford University Press, 1955; London, Collins, 1958.

Cumming, R. D. *The Philosophy of Jean-Paul Sartre.* New York, Random House, 1965.

de Cayeux, A. *Trente millions de siècles de vie*. Paris, André Bonne, 1958.

De Vore, I. (Ed.). *Primate Behavior*. New York, Holt, Rinehart and Winston, 1965.

Dobzhansky, T. "Evolution and Environment." In Sol Tax (Ed.) *Evolution after Darwin*. Vol. 1. Chicago and London, University of Chicago Press, 1960.

—— *Mankind Evolving*. New Haven, Yale University Press, 1963.

Donceel, J. F. "Teilhard de Chardin: Scientist or Philosopher?" *International Philosophical Quarterly*, 5: 248-66, 1965.

Dubos, R. *The Torch of Life*. New York, Simon & Schuster, 1962.

Dunham, B. *Heroes and Heretics*. New York, Knopf, 1964.

Durkheim, E. (1915). "The Elementary Forms of the Religious Life." In W. A. Lessa and E. Z. Vogt (Eds.) *Reader in Comparative Religion*. Evanston, Row Peterson, 1958.

Eccles, J. *The Neurophysiological Basis of Mind*. New York, Oxford University Press; Oxford, Clarendon Press, 1953.

—— *The Brain and the Unity of Conscious Experience*, Cambridge, Cambridge University Press, 1965.

Etkin, W. "Social Behavioral Factors in the Emergence of Man." In S. M. Garn (Ed.) *Culture and the Direction of Human Evolution*. Detroit, Wayne State University Press, 1964.

—— (Ed.) *Social Behavior and Organization among Vertebrates*. Chicago and London, University of Chicago Press, 1964.

Feibleman, J. K. *Mankind Behaving*. Springfield, Charles C. Thomas, 1963.

Fox, S. W. (Ed.). *The Origin of Prebiological Systems*. New York, Academic Press, 1965.

Francoeur, R. T. *Perspectives in Evolution*. Baltimore, Helicon, 1965.

Frankel, C. *The Case for Modern Man*. Boston, Beacon Press, 1959; London, Macmillan, 1957.

Fromm, E. "Value, Psychology, and Human Existence." In A. H. Maslow (Ed.) *New Knowledge of Human Values*. New York, Harper & Row, 1959.

—— *The Heart of Man*. New York, Harper & Row, 1964; London, Routledge, 1965.

Gabor, D. *Inventing the Future*. New York, Knopf; London, Penguin Books, 1964.

Goode, W. J. (1955). "Contemporary Thinking about Primitive Religion." In M. J. O'Leary (Ed.) *Readings in Cultural Anthropology*. New York, Selected Academic Reading Inc., 1965.

Goodenough, E. R. *The Psychology of Religious Experience*. New York, Basic Books, 1965.

Hallowell, A. I. "Self, Society, and Culture in Phylogenetic Perspective." In S. Tax (Ed.) *Evolution after Darwin*, Vol. II. Chicago and London, University of Chicago Press, 1960.

—— "The Protocultural Foundations of Human Adaptation," pp. 236-55. In S. L. Washburn (Ed.) *Social Life of Early Man*. New York, Wenner-Gren Foundation, 1961; London, Methuen, 1962.

Hartshorne, C. *Beyond Humanism*. Chicago, Willet Clark, 1937.

—— *The Logic of Perfection*. Lasalle, Illinois, Open Court, 1962.

Hawkes, J., and Wooley, L. *Prehistory and the Beginning of Civilization. History of Mankind*, Vol. I. New York, Harper & Row, 1963.

Heim, K. *Christian Faith and Natural Science*. New York, Harper & Row; London, SCM Press, 1953.

Henry, J. *Culture against Man*. New York, Random House, 1963.

Herrick, C. J. *The Evolution of Human Nature*. New York, Harper & Row, 1956.

Hockett, C. F. "Animal 'Languages' and Human Language."
In J. N. Spuhler (Ed.) *The Evolution of Man's Capacity
for Culture*. Detroit, Wayne State University Press, 1959.
—— and R. Ascher. "The Human Revolution." *Current
Anthropology*, 5 : 135-68, 1964.

Hoyle, F. *Frontiers of Astronomy*. New York, Harper & Row;
London, Heinemann, 1955.
—— "Recent Developments in Cosmology." *Nature*, 208 : 11-
114; 1965.

Huxley, J. S. "Evolution, Culture and Biology." In W. L.
Thomas (Ed.) *Yearbook of Anthropology, 1955*. New
York, Wenner-Gren Foundation, 1955.

James E. O. *Prehistoric Religion*. New York, Barnes & Noble;
London, Thames & Hudson, 1957.

Kahler, E. *The Meaning of History*. New York, Braziller,
1964; London, Chapman & Hall, 1965.

Kardiner, A. *The Psychological Frontiers of Society*. New
York, Columbia University Press, 1963.

Krutch, J. W. *The Measure of Man*. New York, Grosset &
Dunlap, 1953.

Lack, D. *Evolutionary Theory and Christian Belief*. London,
Methuen, 1957.

Langer, S. K. *Philosophy in a New Key*. New York, Penguin,
1948; London, Oxford University Press, 1952.

Lederberg, J. "Signs of Life. Criterion-system of Exobiology."
Nature, 207 : 9-13, 1965.

Long, C. H. *Alpha: The Myths of Creation*. New York,
Braziller, 1963.

Malinowski, B. (1931). "The Role of Magic and Religion."
In W. A. Lessa and E. Z. Vogt (Eds.) *Reader in Comparative Religion*. Evanston, Row Peterson, 1958.

Malraux, A. *The Metamorphosis of the Gods*. New York,
Doubleday; London, Secker & Warburg, 1960.

Matson, F. W. *The Broken Image*. New York, Braziller, 1964.

Mead, M. (Ed.). *Cultural Patterns and Technical Change*.
New York, New American Library, 1955.

Menaker, E., and Menaker, W. *Ego in Evolution*. New York, Grove Press, 1965.

Munroe, R. L. *Schools of Psychoanalytic Thought*. New York, Holt, Rinehart, 1955; London, Hutchinson, 1957.

Murphy, G. *Human Potentialities*. New York, Basic Books, 1958; London, Allen & Unwin, 1960.

Murray, H. A. "Two Versions of Man." In H. Shapley (Ed.) *Science Ponders Religion*. New York, Appleton-Century-Crofts, 1960.

Nagel, E. *The Structure of Science*. New York, Harcourt Brace, 1963; London, Routledge, 1961.

Oakley, K. P. *Man the Tool-maker*. London, British Museum of Natural History, 1961.

—— *Frameworks for Dating Fossil Man*. Chicago, Aldine, 1964.

Oparin, A. I. (Ed.). "The Origin of Life on Earth." *Reports of the International Symposium*. Moscow, Academy of Science, USSR, 1959.

—— *Life, Its Nature, Origin and Development*. Edinburgh, Oliver & Boyd, 1961.

Penfield, W., and Roberts, L. *Speech and Brain Mechanisms*. Princeton, Princeton University Press, 1959.

Present, I. I. "On the Essence of Life in Connection with its Origins." In *O sushchnosti zhizni*. Moscow, Nauka, 1963.

Radin, P. *Primitive Man as Philosopher*. New York, Dover, 1957; London, Constable, 1958.

Rensch, B. *Evolution Above the Species Level*. New York, Columbia University Press, 1960; London, Methuen, 1959.

Roe, A. "Psychological Definitions of Man." In S. L. Washburn (Ed.) *Classification and Human Evolution*. Chicago, Aldine, 1963; London, Methuen, 1964.

Roe, A., and Simpson, G. G. (Eds.). *Behavior and Evolution*. New Haven, Yale University Press, 1958; London, Oxford University Press, 1959.

Russell, B. *A History of Western Philosophy.* New York, Simon & Schuster; London, Allen & Unwin, 1945.

Schaller, G. *The Mountain Gorilla.* Chicago and London, Chicago University Press, 1963.

Sherrington, C. *Man on His Nature,* Garden City, Doubleday Anchor, 1953; London, Penguin Books, 1955.

Simpson, G. G. *The Major Features of Evolution.* New York, Columbia University Press, 1953; London, Oxford University Press, 1954.

—— *This View of Life.* New York, Harcourt, Brace & World, 1964.

—— "Biological Sciences." In *The Great Ideas Today, 1965.* Encyclopaedia Britannica, 1965.

Sinnott, E. W. *The Biology of the Spirit.* New York, Viking Press, 1955; London, Gollancz, 1956.

—— *Matter, Mind and Man.* New York, Harper, 1957; London, Allen & Unwin, 1937.

Sorokin, P. A. *Social and Cultural Dynamics.* New York, Bedminster Press; London, Allen & Unwin, 1937.

Southwick, Ch. H. *Primate Social Behaviour.* New York, Van Nostrand, 1963.

Sullivan, W. *We Are Not Alone.* New York, McGraw Hill, 1964; London, Hodder & Stoughton, 1965.

Teilhard de Chardin, P. *The Phenomenon of Man.* New York, Harper; London, Collins, 1959.

—— *The Future of Man.* New York, Harper; London, Collins, 1964.

Thorpe, W. H. *Biology and the Nature of Man.* London, Oxford University Press, 1962.

—— *Learning and Instinct of Animals.* 2nd ed. London, Methuen, 1963.

—— *Science, Man and Morals.* London, Methuen, 1965.

Tillich, P. *Theology of Culture.* New York, Oxford University Press, 1959.

—— *Systematic Theology.* Vol. III. Chicago and London, Chicago University Press, 1963.

Toynbee, A. *An Historian's Approach to Religion*. London, Oxford University Press, 1956.

—— *A Study of History*, Vol. XII. *Reconsiderations*. London, Oxford University Press, 1961.

Urey, H. C. *The Planets, Their Origin and Development*. New Haven, Yale University Press; London, Oxford University Press, 1952.

Waddington, C. H. *The Ethical Animal*. London, Allen & Unwin, 1960.

Wald, G. "Phylogeny and Ontogeny at the Molecular Level." In A. I. Oparin (Ed.) *Evolutionary Biochemistry*. London, Pergamon, 1963.

Washburn, S. L. (Ed.). *Social Life of Early Man*. New York, Wenner-Gren Foundation, 1961; London, Methuen, 1962.

—— *Classification and Human Evolution*. New York; Wenner-Gren Foundation, 1963; London, Methuen, 1964.

Weaver, W. "Scientific Explanation." *Science*, 143 : 1297-1300, 1964.

White, A. D. (1895). *A History of the Warfare of Science with Theology in Christendom*. New York, Braziller; London, Arco Publishers, 1955.

White, L. *The Science of Culture*. New York, Grove Press, 1949.

Wright, S. "Biology and the Philosophy of Science." *The Monist*, 48 : 265-90, 1964.

Zenkovsky, V. V. *A History of Russian Philosophy*. New York, Columbia University Press; London, Routledge, 1953.

Index